NANOTECHNOLOGY SCIENCE AND TECHNOLOGY SERIES

# POLYMER NANOCOMPOSITES: ADVANCES IN FILLER SURFACE MODIFICATION TECHNIQUES

[redacted]

# NANOTECHNOLOGY SCIENCE AND TECHNOLOGY SERIES

**Safe Nanotechnology**
*Arthur J. Cornwelle*
2009. ISBN: 978-1-60692-662-8

**National Nanotechnology Initiative: Assessment and Recommendations**
*Jerrod W. Kleike (Editor)*
2009. ISBN 978-1-60692-727-4

**Nanotechnology Research Collection - 2009/2010. DVD edition**
*James N. Ling (Editor)*
2009. ISBN 978-1-60741-293-9

**Nanotechnology Research Collection - 2009/2010. PDF edition**
*James N. Ling (Editor)*
2009. ISBN 978-1-60741-292-2

**Safe Nanotechnology in the Workplace**
*Nathan I. Bialor (Editor)*
2009. ISBN 978-1-60692-679-6

**Strategic Plan for NIOSH Nanotechnology Research and Guidance**
*Martin W. Lang (Author)*
2009. ISBN: 978-1-60692-678-9

**Nanotechnology in the USA: Developments, Policies and Issues**
*Carl H. Jennings (Editor)*
2009. ISBN: 978-1-60692-800-4

**New Nanotechnology Developments**
*Armando Barrañón (Editor)*
2009. ISBN: 978-1-60741-028-7

**Electrospun Nanofibers and Nanotubes Research Advances**
*A. K. Haghi (Editor)*
2009. ISBN: 978-1-60741-220-5

**Electrospun Nanofibers and Nanotubes Research Advances**
*A. K. Haghi (Editor)*
2009. ISBN: 978-1-60876-762-5 (Online Book)

**Nanostructured Materials for Electrochemical Biosensors**
*Umasankar Yogeswaran, S. Ashok Kuma and Shen-Ming Chen*
2009. ISBN: 978-1-60741-706-4

**Magnetic Properties and Applications of Ferromagnetic Microwires with Amorpheous and Nanocrystalline Structure**
*Arcady Zhukov and Valentina Zhukova*
2009. ISBN 978-1-60741-770-5

**Electrospun Nanofibers Research: Recent Developments**
*A.K. Haghi (Editor)*
2009. ISBN 978-1-60741-834-4

**Nanotechnology: Environmental Health and Safety Aspects**
*Phillip S. Terrazas (Editor)*
2009. ISBN: 978-1-60692-808-0

**Nanofibers: Fabrication, Performance, and Applications**
*W. N. Chang (Editor)*
2009. ISBN: 978-1-60741-947-1

**Carbon Nanotubes: A New Alternative for Electrochemical Sensors**
*Gustavo A. Rivas, María D. Rubianes, María L. Pedano, Nancy F. Ferreyra, Guillermina Luque and Silvia A. Miscoria*
2009. ISBN: 978-1-60741-314-1

**Polymer Nanocomposites: Advances in Filler Surface Modification Techniques**
Vikas Mittal (Editor)
2009. ISBN 978-1-60876-125-8

NANOTECHNOLOGY SCIENCE AND TECHNOLOGY SERIES

# POLYMER NANOCOMPOSITES: ADVANCES IN FILLER SURFACE MODIFICATION TECHNIQUES


**VIKAS MITTAL**

**EDITOR**



**Nova Science Publishers, Inc.**

*New York*

For permission to use material from this book please contact us:
Telephone 631-231-7269; Fax 631-231-8175
Web Site: http://www.novapublishers.com



**Library of Congress Cataloging-in-Publication Data**

Polymer nanocomposites : advances in filler surface modification techniques / editors, Vikas Mittal.
p. cm.
Includes bibliographical references and index.
ISBN 978-1-60876-125-8 (hardcover : alk. paper)
1. Fillers (Materials)--Surfaces. 2. Nanotechnology. 3. Composite materials. I. Mittal, Vikas.
TP159.F47P65 2009
668.4'11--dc22

2009025845

*Published by Nova Science Publishers, Inc. ✢ New York*

# Contents

# Preface

Polymer nanocomposites revolutionized the research in this field owing to the tremendous improvement in the composite properties at very low filler volume fraction. The surface modification of the filler, generally layered silicate montmorillonite clay, is required to compatibilize the organic and inorganic phases. The inorganic clay was modified conventionally with alkyl ammonium ions and the exfoliated nanocomposites with polar polymers could be formed where the clay could be dispersed at nanometer scale. During the initial phase of nanocomposite developments, only ammonium ions of fixed chain length were exchanged on the clay surface. However, this technology suffered when polyolefins and other non-polar polymers were used owing to the difficulties in dispersion of polar clay in the hydrophobic matrices. At best, only partially exfoliated composites could be formed by using these ammonium modified clays. To circumvent these limitations, two possible routes have been followed. By polarizing the polymer matrix (e.g. by addition of compatibilizers or surfactants), one can achieve compatibilization between the organic-inorganic phases. However, this technology leads to deterioration of nanocomposite properties even though better delamination is achieved. On the other hand, one can also focus on the more efficient modification of the filler surface so that the residual polarity after modification of the surface with conventional ammonium ions is also eliminated. A number of new clay surface modification techniques have been developed in the recent years which help in the generation of more exfoliated polymer nanocomposites. These techniques do not rely on the ion exchange of fixed chain length ammonium ions, but lead to generation/exchange of long and polydisperse polymer chains. These techniques include grafting of polymers to the clay surface, grafting of polymers from the clay surface, controlled living polymerization from the clay surface, in situ generation of polyolefins from the clay surface and clay surface reactions etc. and form very robust technologies for the complete organophilization of the clay surface. The generation of thick brushes around the clay surface owing to the better surface modification leads to better coverage of the electrostatic forces binding the clay platelets together and also leads to higher basal plane spacing between them. As a result, the modified platelets are more susceptible to exfoliation when compounded with the polymer matrices.

Chapter I - Polymer nanocomposites are organic-inorganic hybrid materials with at least one dimension of the filler phase less than 100 nm. The fillers used generally for the nanocomposites are high aspect ratio platelets with the thickness near to one nanometer. The uniform dispersion of these platelets in the polymer matrix leads to significant improvements in the composite properties at very low filler volume fractions. To make these filler platelets compatible with the polymer materials, their surface modification is required. The surface modifications conventionally used are ammonium modifications which are suitable for the generation of polar nanocomposites, however the non-polar nanocomposites did not have much success. The use of low molecular weight compatibilizer used to achieve compatibilization between the filler phase and the non-polar polymer phase also suffers from various limitations as it negatively impacts the composite properties. The thermal stability of the ammonium modifications is also of concern as their onset of degradation lies in the similar temperature range as required for the compounding of the polymers with the filler. Therefore there is a need to design new thermally stable surface modifications which can remain stable during the temperature range of compounding as well for the entire time scale of compounding process. A small amount of thermal degradation of the filler modification can lead to unwanted side reaction with the polymer matrix leading to disturbances at the organic-inorganic interface as well as reduction the molecular weight of the polymer matrix etc. To achieve the delamination of filler especially in the case of non-polar polymer systems, the approach of surface optimization of the filler is more beneficial approach as it does not require the use of compatibilizer and improvement in the composite properties can still be achieve without the use of compatibilizer. Under this approach, the aim is to enhance the basal plane spacing of the filler as much as possible so that the electrostatic forces of interactions holding the filler platelets together become weak and the platelets can then be exfoliated under the action of shear when mixed at high temperature with the polymer. Increased chain density in the surface modification leads to better results owing to these reasons. Apart from that, ionically bound long chains can be achieved on the filler surface by using more advanced surface modification techniques like polymerization to the surface and polymerization from the surface. Living or controlled polymerization methods can also be used to achieve well controlled brushes of homopolymers or copolymers ionically bound on the clay surface. Apart from that, simple surface reactions can be achieved on the surface to improve the basal plane spacing. Physical adsorption on the filler surface has also been reported which not only increases the basal plane spacing of the filler and hence its susceptibility to exfoliate in the polymer matrix, but also the thermal stability of the organic modification.

Chapter II - Intense research activity in the field of nano composites has gained momentum in the recent years because of their enhanced properties compared to the pure polymers and the conventional fibre filler reinforced composites. In the early 1990s Toyota Researchers discovered a new class of materials with superior properties based on incorporation of nano-sized clay particles into nylon polymer matrix. Since then, many attempts are being made to understand the exfoliation of clay layers in order to incorporate them in the polymer matrix with the basic objective to prepare polymer-clay nanocomposites (PCNs) as advanced materials with much improved mechanical and thermal properties.

Swelling clays such as montmorillonite contain aggregated clay sheets having thickness in the nano range. Exfoliation of the aggregated clay sheets would result in clay sheets of nano thickness which when uniformly distributed in a polymer matrix would result in PCNs. Understanding the various methods to completely exfoliate the clay sheets and distributing the latter uniformly is a challenging task and requires the understanding of the basic work carried out in intercalation and interlayer chemistry involved in organo- clays. Esterification is one of the processes which are readily catalysed by clays which can be very easily modified to generate acid sites. . Preparation of esters including large ones can be carried out on the surface of clays including the interlayer region where suitable ambience could be created through different types of modification. This could be advantageously used to bring about exfoliation. The objective of this chapter is to review the work that has been carried out in the recent years on the different types of surface modifications and other techniques and practices employed to bring about esterification processs on the clay surfaces that possibly could be employed to achieve exfoliation desired for preparation of polymer clay nanocomposites.

Chapter III - Several methods have been used in the preparation of exfoliated clay-polymer nanocomposites, e. g., in-situ polymerization of monomers, melt blending and solution blending in polar solvents. Of all the methods used in the preparation of nanocomposites, in situ polymerization method offers the best control over the polymer architecture and the structure of the composites. Various different polymerization methods were employed in the grafting of polymer chains from clay surface, including living radical polymerization method, living anionic polymerization method, ring-opening polymerization method and coordination polymerization method. The application of the polymerization methods in grafting of polymer chains on clay surface was summarized in this chapter. Besides grafting of polymer chains on clay surface, the grafting of polymer chains on the edge of clay layers was also reviewed.

Chapter IV - Grafting of polymer chains from the clay surface are made it successful with the help of cationic azo-initiators containing quaternary ammonium functional groups. Monocationic and bicationic initiators are designed for this purpose and a simple ion-exchange method is used to prepare initiator-bound clay layers. The interlayer *d*-spacing of the resulting initiator-bound clays are largely dependent on the size and their orientation of the organic initiators. The competence between the monocationic and bicationic initiators toward the surface functionalization upon ion exchange is discussed. The chain length and diameter of initiator molecules are facilities a higher *d* spacing in bicationic initiator-bound clay than the monocationic. The successful exfoliation of polymer clay nanocomposites resulted from the polymer grafting from the clay surface makes superiority over the other techniques. Hence, they offered a very high improvement in both thermal and mechanical properties, due to the high aspect ratio of individual clay layers that are uniformly dispersed throughout the polymer matrix.

Chapter V - This chapter summarizes the use of atom transfer radical polymerization (ATRP) for the preparation of polymer-layered silicate nanocomposites (PLSNs). The ATRP mechanism does not appear to be hampered by the presence of the clay, resulting in well controlled and predictable molecular weight distributions. Initiation of ATRP can be accomplished using intercalated alkyl halides, and results in polymer chains that are grafted from the silicate layers. Production of various copolymer types, such as block, gradient, graft and statistical copolymers, can be achieved. These exhibit interesting phase separation and switchable surfaces if treated with selective solvents. The resultant nanocomposites structures are, like most in situ preparation methods, often a mixture of exfoliated and intercalated structures are formed. However, given the flexibility of ATRP in making a range of polymer architectures and functionalities, it is expected that enhanced exfoliation in PLSNs can be achieved by grafting polymers from the silicate surface using ATRP.

Chapter VI - In this chapter, the authors have discussed the cationic exchange process which is the ordinary procedure to adsorb a chemical containing a charge-bearing group (i.e. an ionized molecule) on the internal surface of clay galleries. If the adsorbate is an organic molecule containing a charge-bearing group with a high affinity for clay immobilized anions (e.g. a quaternary ammonium salt), the cationic exchange process modifies clay nature (from hydrophilic to hydrophobic) to an extent which depends on the nature of the organic molecule. This treatment generates organically-modified clays, characterized by an improved compatibility with most of the polymers. If the adsorbate bears more than one function (e.g. cationic fluorescent dyes contain charge-bearing any fluorophore-bearing groups), cationic exchange process is a suitable method to render clays photo-active. Photofunctionalization can open large possibilities for all the modern on-line process monitoring techniques by conceiving a detection system capable of perform real-time monitoring of melt compounding of a polymer and a lamellar filler and sensitive to multiscale morphologies.

Chapter VII - This article reviews the essential feature of clay minerals, the possibilities of the clay modification and their use as potential adsorbents for the environmental remediation. The structure of clays is based on extensive sheets of silicon bonded with oxygen combined with metal oxide flat sheets (usually Al or Mg). Polar molecules are able to force the silicate layers apart because of relatively weak bonding. The swelling of clay minerals is an example of its electrical double layer properties. Properties of crystalline bonding and symmetry affect the charge on the crystal and affect the layer ratio of cations to anions. Cation exchange capacity is one of the most important characteristics of clay particles. Clays can be modified by metals, surface active agents (surfactants) or a mixture of the two. Montmorillonite and their modified products were assessed for the adsorption performance in the treatment of metals (e.g. Cu) and organic matter (e.g., phenol) in water. The results are discussed in the associated sections with the surface properties and structures of various clays.

Chapter VIII - Polymer nanocomposites require the compatibilization of the organic and inorganic phases in order to achieve optimal microstructure of the composites and subsequently improved composite properties owing to the delamination of the filler. Therefore, the surface modification of the filler surface is achieved which makes it compatible with the polymer matrices. The nature of the achieved modified filler surface as well as its cleanliness has tremendous impact on the composite properties. The cleanliness of the filler surface is analyzed by using high resolution thermogravimetric analysis where the filler modification molecules not bound to the filler surface and present only as pseudo bilayer degrade at lower temperature and the absence of low temperature degradation peak in the thermogravimetric analysis thus would confirm the achievement of clean filler surface. The X-ray diffraction is not suitable to ascertain the cleanliness of the surface as the clean fillers as well as the fillers with excess modification have been observed to have similar basal plane spacing values. The removal of the excess modification is necessary as it may interact with the polymer matrix in more than one ways. At high temperature, its low temperature degradation can impact the thermal properties of the composites, it may also have unwanted chemical reaction with the polymer matrix thus degrading the polymer molecular weight, it may also gelate the dispersion of the filler with polymer in solvents thus destabilizing the system. The presence of excess amount of surface modification can also lead to increased free volume at the interface owing to the non-optimum development of the interface where the organic phase may not be fully compatible with the surface modification. The commercially available modified montmorillonites have been conformed confirmed to contain an excess of surface modification molecules, therefore, their use as it is may have additional interactions in the composites. The presence of the excess of the surface modification molecules on the surface was found to have significant effect on the composite properties, both mechanical and permeation properties were observed to decrease as a result of the excess surface modification. The microstructure as studied by transmission electron microscope however did nit not show any difference in the systems with or no excess thus indicating that the developed microstructure may not be affected by the presence of these excess ions, but these ions surely change the interfacial interactions leading to deterioration in the composite properties.

Chapter IX - Carbon nanotubes, owing to their remarkable mechanical, thermal and electronic properties, have high potential of application in a number of areas as biosensing devices, field effect transistors, chemical sensors, high performance nanocomposites and other nanoscale devices and circuits. These applications require the nanoscale dispersion of nanotubes in the polymer matrices or various solvents. However, nanotubes being inert at surface do not swell in most of the solvents and polymers in their pristine state thus hindering their large scale application. Tremendous research efforts have been undertaken to achieve compatibility of the nanotubes with these media so as to achieve the nanoscale dispersion of the nanotubes which would translate into significantly better and novel properties than any of the existing systems. Two different approaches of covalent and noncovalent modification of nanotubes have been reported to achieve such compatibility. Noncovalent functionalization is based on a number of phenomena e.g. wrapping of polymer chains around the nanotubes, physical adsorption of polymer chains or surfactants on the nanotube surface and deposition of functionalizing molecules by pi-pi stacking etc. Apart from that, charged inorganic particles have been used to achieve the swelling of nanotubes. Generation of thin films of silica or various polymers from the non-covalently bound initiators or monomers have also been reported. The chemical groups generated during the functionalization process were also reacted with proteins to immobilize them on the nanotube surfaces. Noncovalent functionalization of nanotubes leads to nanotube modifications without any loss in the physical properties of nanotubes, however the strength of the modification attached to the nanotubes for load transfer may not be enough. Covalent functionalization has been reported to generate more stronger attachment of the modification on the nanotube surface and thus can be more efficient in load transfer applications. However, the covalent treatment of nanotubes may generate defects in their walls leading to reduction in the mechanical and conducting properties of nanotubes. A large number of methods have been reported for covalent modification of nanotubes e.g. attachment of polyethyleneimine, polyethylene glycol or proteins with amino end groups to the nanotubes, grafting of polymers to the nanotube surface, grafting of polymers from the nanotube surface and plasma treatment for polymer grafting etc. Covalent binding of inorganic nanoparticles to the nanotube surface has also been reported. Use of carbanions to open the pi bonds on nanotube surfaces and subsequent initiation of polymerization from the surface has also been successful. Moreover, various covalent functionalizing techniques avoiding the use of harsh acid treatment of nanotubes have been developed.

Chapter X - This chapter discusses a unique chain-end functionalized polyolefin structure that contains a reactive terminal group for forming a strong interaction with clay surfaces, resulting in exfoliated polyolefin/clay nanocomposites. Two commercially important polymers, including isotactic polypropylene (PP) and poly(vinylidene fluoride) (PVDF), have been studied to illustrate the research idea and experimental results. The chemistry to prepare chain-end functionalized PP containing a terminal OH or $NH_3^+Cl^-$ group involves metallocene-mediated propylene polymerization using rac-$Me_2Si[2\text{-}Me\text{-}4\text{-}Ph(Ind)]_2ZrCl_2$/MAO complex in the presence of styrene derivatives (St-f) and hydrogen, which serve as the chain transfer agents. On the other hand, the chain end functionalized fluoropolymer is prepared by a specific borane control radical initiator that bears a reactive $Si(OR)_3$ group that is directly transferred to the beginning of the polymer chain. Despite the extremely low concentration of the chain-end group in PP and PVDF polymers, the terminal OH, $NH_3^+Cl^-$, and $Si(OR)_3$ groups show good accessibility and reactivity toward the clay surfaces and form hydrogen bond, ion-exchange the cations ($Li^+$, $Na^+$, etc.), and chemical bonding, respectively. Such strong interactions anchor the PP or PVDF chain to the clay surfaces. On the other hand, the remaining unperturbed end-tethered high molecular weight PP or PVDF tail exfoliates the clay layers. This exfoliated structure is maintained even after further mixing with pure neat PP or PVDF polymers.

In: Polymer Nanocomposites: Advances in Filler Surface
Editor: Vikas Mittal
ISBN: 978-1-60876-125-8


*Chapter I*

# Need of New Surface Modifications

***Vikas Mittal****
Department of Chemistry and Applied Biosciences, Institute of Chemical and Bio Engineering, Swiss Federal Institute of Technology, ETH Hoenggerberg, 8093, Zurich, Switzerland

## Abstract

Polymer nanocomposites are organic-inorganic hybrid materials with at least one dimension of the filler phase less than 100 nm. The fillers used generally for the nanocomposites are high aspect ratio platelets with the thickness near to one nanometer. The uniform dispersion of these platelets in the polymer matrix leads to significant improvements in the composite properties at very low filler volume fractions. To make these filler platelets compatible with the polymer materials, their surface modification is required. The surface modifications conventionally used are ammonium modifications which are suitable for the generation of polar nanocomposites, however the non-polar nanocomposites did not have much success. The use of low molecular weight compatibilizer used to achieve compatibilization between the filler phase and the non-polar polymer phase also suffers from various limitations as it negatively impacts the composite properties. The thermal stability of the ammonium modifications is also of concern as their onset of degradation lies in the similar temperature range as required for the compounding of the polymers with the filler. Therefore there is a need to design new thermally stable surface modifications which can remain stable during the temperature range of compounding as well for the entire time scale of compounding process. A small amount of thermal degradation of the filler modification can lead to unwanted side reaction with the polymer matrix leading to disturbances at the organic-inorganic interface as well as reduction the molecular weight of the polymer matrix etc. To achieve the delamination of filler especially in the case of non-polar polymer systems, the approach of surface optimization of the filler is more beneficial approach as it does not require the use of compatibilizer and improvement in the composite properties can still be achieve without the use of compatibilizer. Under this approach, the aim is to enhance

* Corresponding author: BASF SE, Polymer Research, Ludwigshafen, Germany vikas.mittal@basf.com

the basal plane spacing of the filler as much as possible so that the electrostatic forces of interactions holding the filler platelets together become weak and the platelets can then be exfoliated under the action of shear when mixed at high temperature with the polymer. Increased chain density in the surface modification leads to better results owing to these reasons. Apart from that, ionically bound long chains can be achieved on the filler surface by using more advanced surface modification techniques like polymerization to the surface and polymerization from the surface. Living or controlled polymerization methods can also be used to achieve well controlled brushes of homopolymers or copolymers ionically bound on the clay surface. Apart from that, simple surface reactions can be achieved on the surface to improve the basal plane spacing. Physical adsorption on the filler surface has also been reported which not only increases the basal plane spacing of the filler and hence its susceptibility to exfoliate in the polymer matrix, but also the thermal stability of the organic modification.

**Keywords**: organic-inorganic hybrid materials, aspect ratio, platelets, ammonium modifications, interface, delamination of filler, compatibilizer, composite properties, thermal stability of modification.

## 1.1. Introduction

Polymer nanocomposites use the layered silicates aluminosilicate platelets as fillers in order to achieve improvements in the properties as compared to the pure polymer. These aluminosilicates, of which montmorillonite is most commonly used, are polar in nature with alkali and alkaline earth metal cations adsorbed on the surface to compensate the negative charges in the crystal structure. Owing to the electrostatic forces of attraction, these filler platelets form thick stacks thus forming the interlayers in between the platelets. To make these fillers compatible with most of the polymer matrices and hence to delaminate them fully into their nanometer thick platelets in the polymer matrices, the surface modification is necessary. It is achieved by the exchange of the alkali and alkaline earth metal cations on the surface with long chain alkyl ammonium cations [1-5]. These cations bring higher basal plane or interlayer spacing between the individual platelets and reduce the surface energy, however, the platelets are still held together by the residual electrostatic forces. The nanocomposites in which these filler materials are delaminated into the nanometer thick primary platelets to provide high aspect ratio, tremendous increase in the properties is generally achieved [6-13]. Owing to the small amount of filler required to achieve the optimum improvement in the composite properties as compared to the conventional microcomposites, the nanocomposites retain the optical clarity and also have similar density as the polymer material itself.

Polymer nanocomposites have been tried with practically all the different polymer matrices with different extents of the exfoliation of the filler in these matrices depending on the nature of the polymer, filler surface modifications, interactions at the organic-inorganic interface and processing conditions. Polar nanocomposites have been more successful in terms of filler exfoliation owing to the better match of polarity of the polar polymer with the partially polar alkyl ammonium ions modified layered silicates [14-19]. The layered silicates are though organically modified, but owing to the lower area of cross-section of the ammonium ions in comparison to the area per charge available on the surface of the montmorillonite platelets, there are still residual forces of interaction present in between the filler platelets leading to the partial polarity of the filler interlayers. The polar polymer chains can thus intercalate in the filler interlayers and during this process can push these platelets to larger distances thus leading to their exfoliation. However, the behavior of non-polar polymer matrices towards the filler delamination is different as compared to the polar polymer systems owing to the absence of any positive interfacial interactions between the organic and inorganic phases. Increasing the length of the tethered surfactants was predicted to improve the thermodynamic state of the system as more distance generated among the clay layers helped in bringing down the effective interactions between the clay sheets. For a given density of alkyl chains on the surface, long chains were predicted to form a more homogenous phase than the short ones [20].

## 1.2. Conventional Polymer Nanocomposite Systems

Figure 1 shows the X-ray diffractograms of the polyurethane, epoxy and polypropylene nanocomposites synthesized by using 3 vol% montmorillonite modified with benzylhexadecyltrimethylammonium modification. There are diffraction peaks present in all the three nanocomposite systems indicating that the filler may not have been fully exfoliated. However, X-ray peaks are very qualitative in nature and are influenced by the sample preparation techniques, orientation of the filler particles and defects in the crystal structure of the inorganic materials. However, as the similar montmorillonites and similar methods of preparation were used, therefore one can qualitatively observe that the polypropylene composite is only intercalated as compared to more exfoliated polar polyurethane and epoxy composites. Also the basal plane spacing in the polypropylene composite is much lower than the polar polymer counterparts. It obviously underlines the phenomenon mentioned earlier that the intercalation of the non-polar polymer chains in the interlayers is challenging especially when the conventional ammonium ions are used as surface modifications indicating that to achieve better exfoliation of the filler, new developments in the system are required. Figure 2 also shows the TEM micrographs of the polymer nanocomposites systems mentioned above and it is visible from these TEM micrographs that mostly a mixture of intercalated and exfoliated morphologies is obtained in the composite microstructure. Though the polar polymers may have higher extents of filler exfoliation, even in majority of these cases complete exfoliation of the filler is not achieved. Therefore, to achieve much better delamination of the filler in all the composite systems, more improvements in the system are required especially in the case of non-polar polymers like polyolefins.

A common approach to circumvent the problem of hydrophobicity of the polyolefin matrices is the partial polarization of the matrix by the addition of low molecular weight compatibilizers which owing to their amphiphilic nature help to compatibilize the organic and inorganic phases [21-26]. The compatibilizer though helps to improve the mechanical properties to a certain extent, however also plasticizes the matrix thus leading to the reduction of the modulus after a certain weight fraction. This is owing to the competitive forces of modulus increase owing to filler exfoliation and modulus decrease owing to matrix plasticization. Also, the addition of compatibilizer can negatively impact the crystallinity and barrier properties of the materials. To achieve a meaningful exfoliation of the filler in the polymer matrix, large amount of compatibilizer is required which is further detrimental to the polymer properties. Also, it has been observed that compatibilizer chains intercalated the clay interlayers leaving the high molecular weight polymer chains behind. Figure 3 shows the effect of increasing amount of compatibilizer on the tensile modulus of the polypropylene nanocomposites with 3 vol% of the filler modified with dioctadecyldimethylammonium. The modulus increases first owing to the exfoliation of the filler, but starts to decrease after 2 wt% owing to above mentioned matrix plasticization overcoming the property improvements due to filler exfoliation [27]. It was also observed that the barrier properties of the composites were also either unaffected or were negatively impacted by the addition of small amounts of compatibilizer and the incompatibility between the non-polar filler surface modification and polar compatibilizer chains was observed to cause this effect. Therefore, it is generally clear that the use of compatibilizer does not bring the required goal of increased exfoliation and subsequent corresponding improvement in the polymer properties.

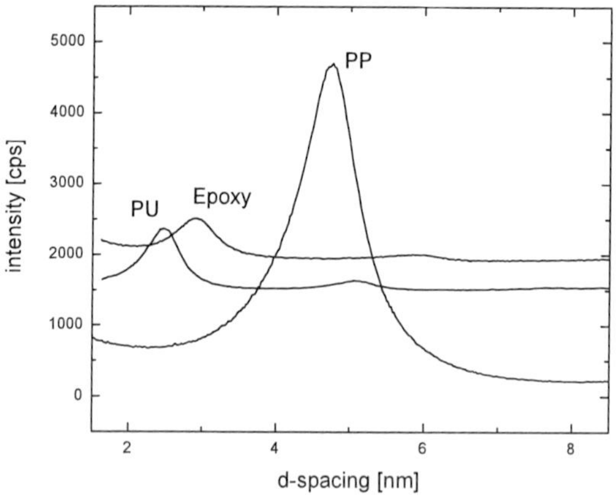


Figure 1. X-ray diffractograms of the polyurethane, epoxy and polypropylene nanocomposites synthesized by using 3 vol% montmorillonite modified with benzylhexadecyltrimethylammonium.

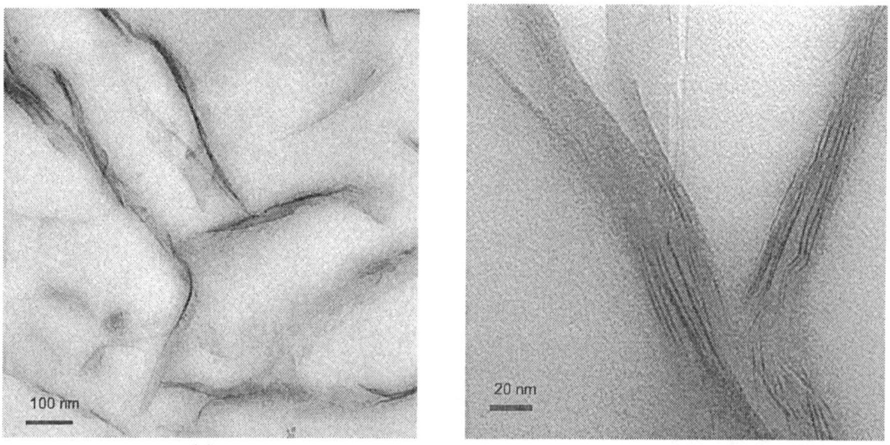


(a) (b)

Figure 2. TEM micrographs of the polymer nanocomposite materials for (a) polypropylene and (b) epoxy polymer matrices.

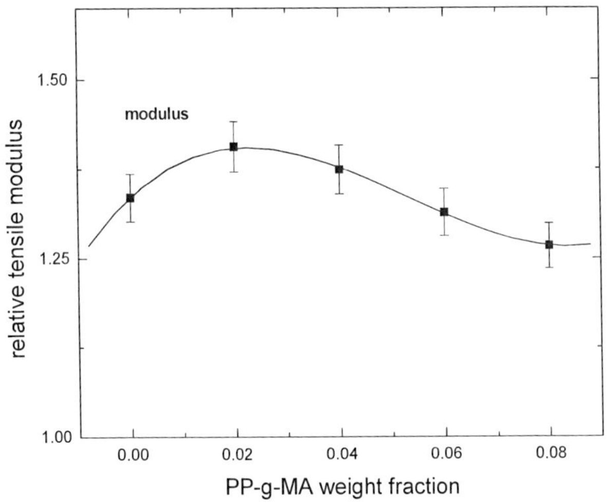


Figure 3. Relative tensile modulus of the polypropylene nanocomposites as a function of the amount of PP-graft-maleic anhydride compatibilizer.

## 1.3. Unconventional Surface Modifications

As the use of partial polarization of the polymer matrix by the addition of compatibilizer does not lead to the achievement of the required goal in the case of non-polar polymer nanocomposites, therefore it is required to develop new approaches to achieve the goal of uniform nanoscale dispersion of the filler platelets. Optimization of the filler surface modification has been developed in the recent years to circumvent the limitations of the compatibilizers. The optimization includes the improvement in the basal plane spacing of the filler to as high value as possible. It has been mentioned in the earlier section that for the polar polymers, it is more the interfacial interactions which are of more importance than the basal plane spacing. However, in the case of non-polar polymer matrices, there is no interaction present at the interface and in such cases, increased basal plane spacing of the filler owing to advanced modification can be very beneficial for its exfoliation in the polymer matrix. Increased basal spacing leads to reduced forces of attraction holding the platelets together and thus these loosely held platelets are more susceptible to exfoliation when sheared in the compounder along with the polymer matrix. The most obvious way to achieve higher basal plane spacing in the modified fillers is by increasing the chain density in the ammonium modifications. By increasing the chain density of the ammonium modifications, the cross-sectional area of the ammonium group increases. Thus it occupies more space on the filler surface lading to more crowding of the surface modification on the filler surface which subsequently leads to more vertical positioning of the surface modification molecules than the case when the chain density in the modification molecules is low. This straighter positioning of the modification molecules leads to higher basal plane spacing and subsequent reduction in the electrostatic forces of interaction holding the platelets together. Figure 4 shows the filler platelets modified with trioctadecylmethylammonium and tetraoctadecylammonium modifications. These surface modifications increase the basal plane spacing of the filler materials to more than 3 nm as compared to 0.9 or 1 nm for the unmodified platelets [28,29]. If higher cation exchange capacity montmorillonite is used, the basal plane spacing can be further increased. The high cation exchange capacity filler has higher charge density on the filler surface which means more amount of organic matter can be ionically exchanged on the clay surface owing to large number of charges per unit area. This then leads to increase in the basal plane spacing and subsequently filler delamination is improved. The polyethylene and polypropylene nanocomposites were synthesized with the montmorillonites modified as in Figure 4 and partial exfoliation of the filler could be achieved even without the use of compatibilizer [28,29]. The composite properties were also significantly improved.

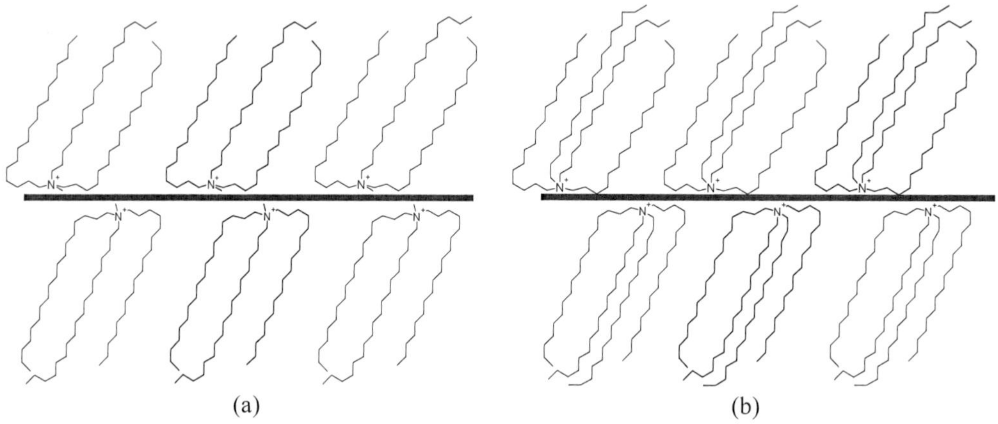

Figure 4. Filler platelets modified with (a) trioctadecylmethylammonium and (b) tetraoctadecylammonium.

Another important point to note that in order to achieve better optimization of the modified filler surface, one should avoid the thermal degradation of the surface modification at the high compounding temperatures. The ammonium modification has onset of degradation in the same temperature range used in the compounding operations, thus requiring caution during compounding as even a small percent of thermal degradation of the ammonium modification can lead to unwanted side reactions with the polymer matrix leading to disturbance at the interface or lowering of the molecular weight of the polymer matrix etc. Therefore, compounding temperature should be kept low in order to avoid the thermal degradation but high enough to ensure uniform mixing of the filler with the polymer. Other possibility is to use the more thermally stable modifications which have higher onset and peak degradation temperatures as compared to ammonium modifications [30]. Figure 5 shows an example of imidazolium modification exchanged on the clay surface. Imidazolium molecules have been proven to be much better thermally stable as shown in the thermogravimetric analysis (TGA) of Figure 6. The onset of degradation of the imidazolium modification on the filler surface was observed to be higher than the ammonium modification by almost 50°C indicating much better thermal stability. The dynamic TGA analysis of the montmorillonites modified with ammonium and imidazolium modifications also showed more thermal degradation of the ammonium modification as compared to the imidazolium counterpart. As mentioned above, degradation of even a few percent of the modification can be detrimental to the composite properties as the degradation causes the rupture of C-N bonds of the ammonium groups. Much better oxygen permeation properties were observed in the case of polypropylene nanocomposites with imidazolium modified montmorillonites as compared to the ammonium system. The mechanical properties of the polypropylene nanocomposites with ammonium modified montmorillonite were, however, a little better than the imidazolium system. It is probably because of the lower basal plane spacing of the imidazolium modified filler. The barrier properties were better as they are more sensitive to the interface compatibility and the decreased thermal degradation of the filler would have helped to build a better interface, but in overall, owing to the lower basal plane spacing of the filler, the exfoliation and hence the composite properties may have been negatively affected, it is therefore obvious that if longer chains were attached the clay surface, much better extents of exfoliation as well as composite properties can be achieved.

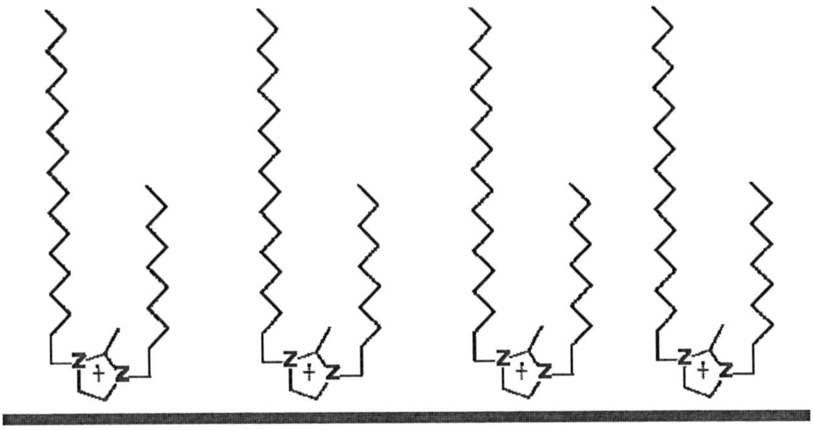

Figure 5. Montmorillonite surface modified with thermally stable imidazolium modification.

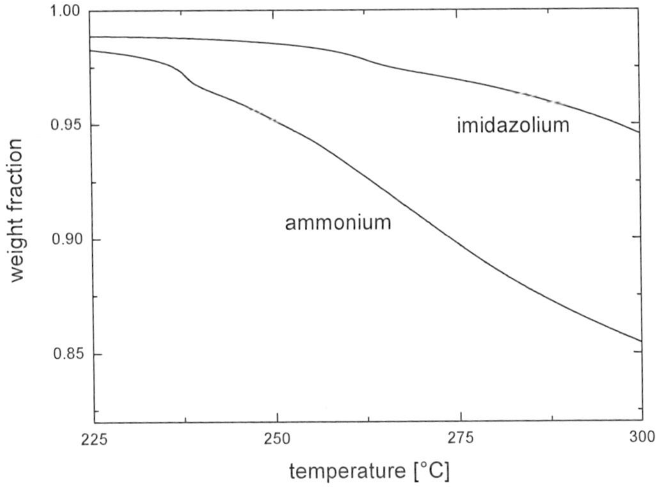


Figure 6. TGA comparison of the montmorillonites modified with ammonium and imidazolium modifications.

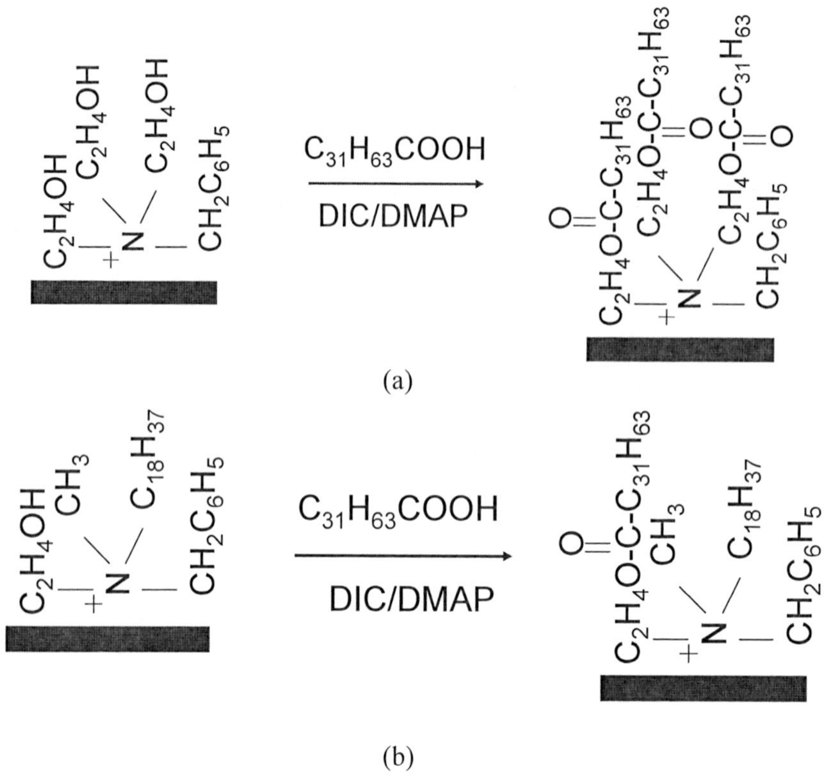


Figure 7. Representation of possibilities of surface esterification reactions on the clay surface [31].

## 1.4. Modified Fillers with Long Chains Attached on the Surface

As mentioned in the section above that the long chains ionically bound to the clay surface as modification would lead to better exfoliation of the filler in the polymer matrix, however, attachment of pre-formed long chains on the filler surface suffers from the limitation of steric hindrance as well as solubility concerns. Therefore, different approach is required to achieve the long chains bound chemically to the filler surface. One such approach to achieve this goal is surface reactions in general and surface esterification reactions in particular [31]. In this approach as shown in Figure 7, a reactive surface modification is first exchanged on the clay surface which is then reacted from outside using simple chemical reactions to bind the long chains to the surface. This way, much longer chains can be attached to the surface without facing the steric hindrance and solubility limitations. As the chain length of external reactant as well as the reactive modification is fixed, therefore the formed brushes of the surface modification on the surface after surface modification have still fixed chain length. As shown in Figure 7, various possibilities of surface esterification reactions are possible owing to the different surface modifications attached to the clay surface. Modifications with varying number of the reactive groups can be used in order to enhance the amount of surface reaction and thus to enhance the amount of organic matter bound to the clay surface. However, one must be careful not to create too much polarity in the interlayer which would then make the intercalation of the polyolefin matrix polymer chains in the interlayer difficult thus nullifying the positive effect of higher basal plane spacing after surface reaction. The basal plane spacing after surface reaction was significantly improved [31]. In one case, the basal plane spacing value of 5.3 nm could be achieved as compared to 1.8 nm for the modified montmorillonite before the esterification reaction.

Among other methods to achieve high basal plane spacing in the filler is the polymerization on the surface approach. Two variations of polymerization on the surface have been developed: polymerization to the surface and polymerization from the surface [32-39]. In the polymerization to the surface approach, a monomer is generally ionically bound to the filler surface which is then copolymerized with an external monomer in the presence of an initiator. This way though a large amount of polymer is formed unattached to the filler surface, but a significant amount of polymer chains could still be attached to the filler surface. In any case, it is only the chain length similar to the size of oligomers which is required to be attached to the surface so that it is long enough to reduce the electrostatic forces of interaction between the clay platelets but is not too long to cause compatibility problems with the polymer matrices when compounded with them (Figure 8).

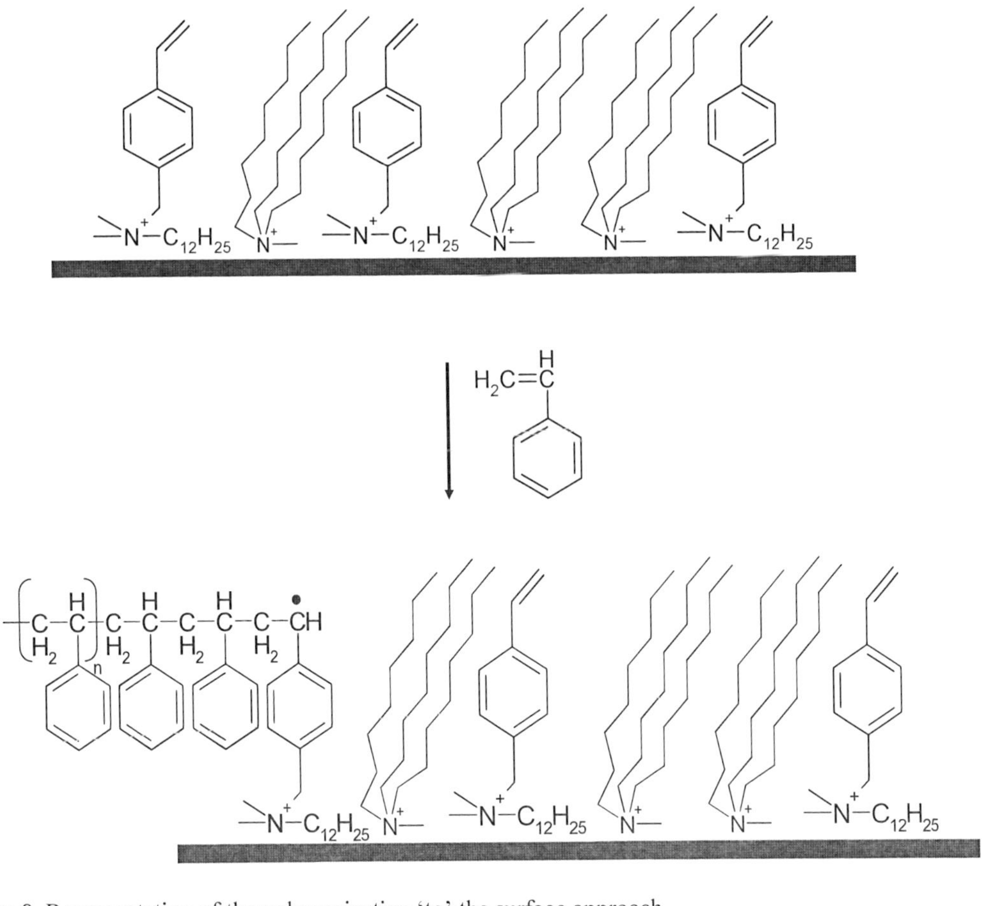


Figure 8. Representation of the polymerization 'to' the surface approach.

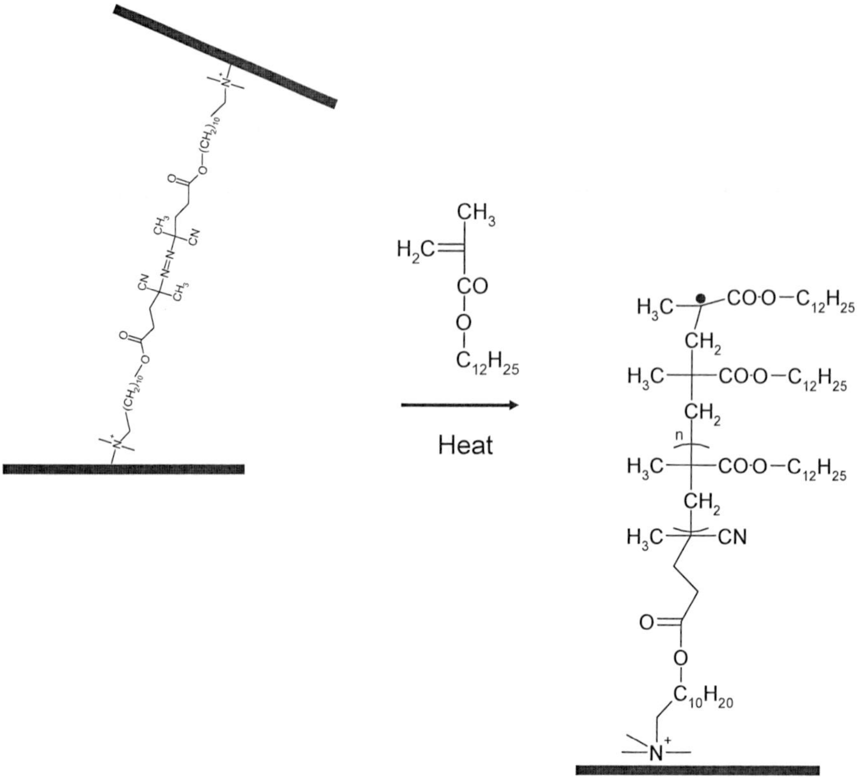


Figure 9. Representation of the polymerization 'from' the surface approach.

One can also partially cover the surface of the filler with the monomer and the rest being with the un-reactive alkyl ammonium chains and thus the amount of the polymerization can also be controlled apart from the control on amount of monomers, polymerization time and temperature etc. Polymerization from the surface is another common approach in which in initiator is ionically bound on the clay surface and these initiator molecules can be used to initiate the polymerization of externally added monomer (Figure 9). This approach leads to much better growth of brushes from the surface of the filler. As seen in the case of polymerization to the surface, the filler surface in this case also be partially modified with the initiator. If the initiator is used to fully cover the surface of the filler, then during the polymerization, the radicals generated from the initiator or growing chain radicals are in very close proximity to each other which subsequently can lead to fast termination. Therefore, it is always beneficial to partially cover the filler surface with the initiator and thus exchanging the non-reactive alkyl ammonium chains in between. There is also a possibility to exchange either monocationic as well as bicationic initiators on the surface [38,39]. In the monocationic initiator, one end of the molecule is ionically bound to the filler surface, whereas the other is free, whereas in the case of bicationic initiator, both the ends of the molecule are ionically bound to the filler surface. The monocationic initiator leads to the generation of one radical bound ionically to the surface whereas the other is let free in solution. The generated radicals are thus not near to each other and have less chances of terminating each other. On the other hand, the free radical in the solution leads to the generation of large amount of free polymer unbound to the filler surface. The bicationic initiator on the other hand, does not lead to higher amounts of free polymer chains in solution, but suffers from faster termination of the radicals as the generated radicals are in close proximity to each other and owing to their immobilization on the filler surface cannot move away from each other faster. The immobilization of the both ends of initiator on different filler platelets can also bridge the clay layers together thus complicating the exfoliation of the filler platelets. Thus both the monocationic as well as bicationic initiators have their positives and limitations and one should carefully choose the nature of initiator to be bound on the clay surface depending on the specific system at hand.

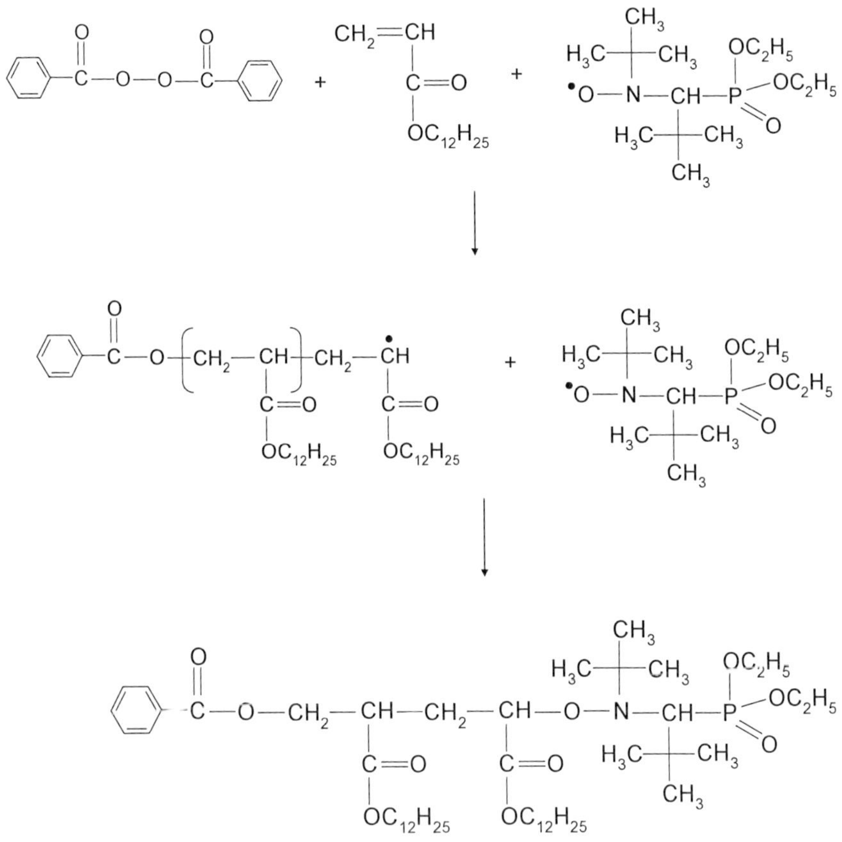

Figure 10. Mechanism of controlled radical polymerization using nitroxide mediated polymerization approach.

Apart from the free radical polymerization approach to achieve polymerization on the clay surface, one can also use the more advanced techniques of controlled radical polymerization to achieve more controlled grafts on the filler surface. There have been a number of developments in this field and examples of two such cases of controlled polymerization are shown in Figures 10 and 11 [40-43]. The mechanism shown are for general nitroxide mediated polymerization and atom transfer radical polymerization, but these systems can be easily adapted to the polymerization on the clay surface leading to controlled polymer brushes ionically bound to the filler surface. Also, these methods allow one to generate not only homopolymers, but also block copolymer thus helping to bring about the compatibility with the polymer matrices.

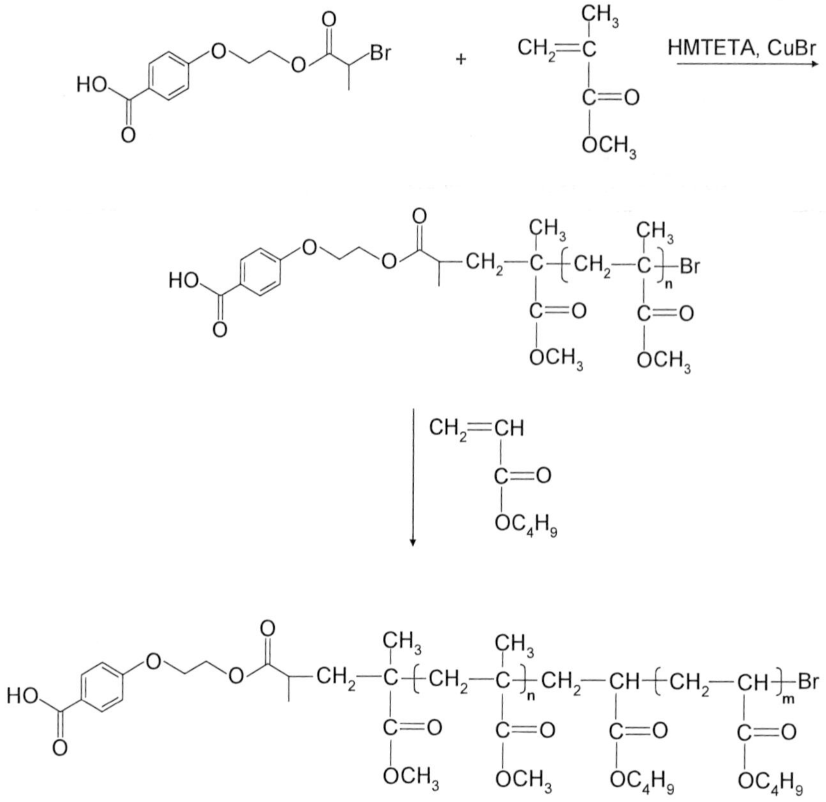


Figure 11. Mechanism of controlled radical polymerization using atom transfer radical polymerization approach.

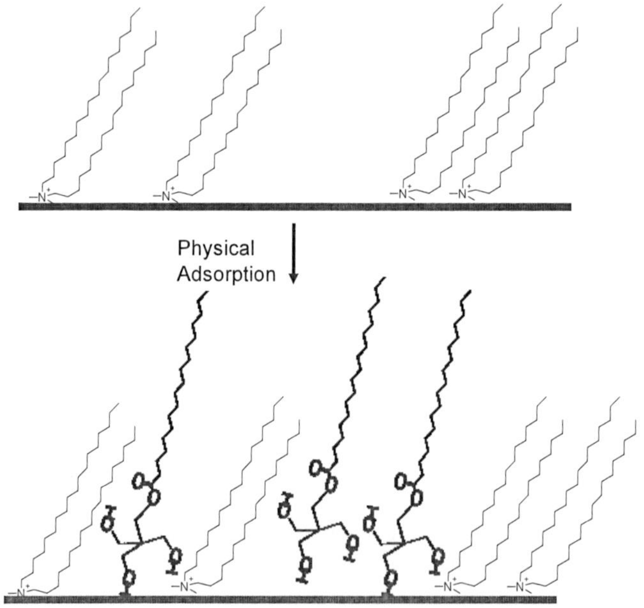


Figure 12. Representation of the physical adsorption process on the surface modified montmorillonite.

There is also the possibility to physically adsorb the organic molecules on the clay surface in order to fully organophilize the clay surface as well as to increase the basal plane spacing. The cross-sectional area of the cation exchanged on the clays surface is generally smaller than the area available per cation on the surface thus leading to spaces on the filler surface in between the surface modifications. These voids as well the edges of the filler platelets can be used to physically adsorb organic molecules by forming hydrogen bonds so that the clay surface is fully covered thus reducing the electrostatic forces of interaction [44]. It was observed that after adsorption of the polymer chains on the surface of modified montmorillonites, the thermal stability of the surface modification was further enhanced. The amount of adsorption could also be optimized as a function of amount of organic molecules or polymer chains initially added for adsorption. Figure 12 is the representation of this phenomenon.

## References


[1] Bailey, SW. In *Reviews in Mineralogy*; Bailey, SW.; Ed.; Virginia Polytechnic Institute and State University: Blacksburg, 1984.

[2] Bailey, SW. In *Crystal Structures of Clay Minerals and their X-ray Identification*; Brindley, GW. Brown, G. Eds.; Mineralogical Society: London, 1980.

[3] Theng, BK. G. *The Chemistry of Clay-Organic Reactions*; Adam Hilger: London, 1974.

[4] Pinnavaia, TJ. *Science* 1983, 220, 365.

[5] Lagaly, G. In *Developments in Ionic Polymers*; Wilson, AD; Prosser, HJ; Eds.; Elsevier Applied Science Publishers: London/New York, 1986.

[6] Lan, T; Kaviratna, PD; Pinnavaia, TJ. *Chem. Mater.* 1994, 6, 573.

[7] Lan, T; Pinnavaia, TJ. *Chem. Mater.* 1994, 6, 2216.

[8] Osman, MA; Mittal, V; Morbidelli, M; Suter, UW. *Macromolecules* 2004, 37, 7250.

[9] Messersmith, PB; Giannelis, EP. *J. Polym. Sci., Part A: Polym. Chem.* 1995, 33, 1047.

[10] Burnside, SD.; Giannelis, EP. *Chem. Mater.* 1995, 7, 1597.

[11] Wang, Z; Pinnavaia, TJ. *Chem. Mater.* 1998, *10*, 3769.

[12] Osman, MA; Mittal, V; Morbidelli, M; Suter, UW. *Macromolecules* 2003, 36, 9851.

[13] Yano, K; Usuki, A; Okada, A. *J. Polym. Sci., Part A: Polym. Chem.* 1997, 35, 2289.

[14] Zilg, C; Thomann, R; Mülhaupt, R; Finter, J. *Adv. Mater.* 1999, 11, 49.

[15] Wang, Z; Pinnavaia, TJ. *Chem. Mater.* 1998, 10, 3769.

[16] Chen, TK; Tien, YI; Wei, KH. *Polymer* 2000, 41, 1345.

[17] Zilg, C; Mülhaupt, R; Finter, J. *Macromol. Chem. Phys.* 1999, 200, 661.

[18] Brown, JM; Curliss, D; Vaia, RA. *Chem. Mater.* 2000, 12, 3376.

[19] Kornmann, X; Thomann, R; Mülhaupt, R; Finter, J; Berglund, L. *J. Appl. Polym. Sci.* 2002, 86, 2643.

[20] Heinz, H; Suter, UW. *Angewandte Chemie Int. Ed.* 2004, 43, 2239.

[21] Kawasumi, M; Hasegawa, N; Kato, M; Usuki, A; Okada, A. *Macromolecules* 1997, 30, 6333.

[22] Usuki, A; Kato, M; Okada, A; Kurauchi, T. *J. Appl. Polym. Sci.* 1997, 63, 137.
[23] Hasegawa, N; Kawasumi, M; Kato, M; Usuki, A; Okada, A. *J. Appl. Polym. Sci.* 1998, 67, 87.
[24] Reichert, P; Nitz, H; Klinke, S; Brandsch, R; Thomann, R; Mülhaupt, R. *Macromol. Mater. Eng.* 2000, 275, 8.
[25] Manias, E; Touny, A; Wu, L; Strawhecker, K; Lu, B; Chung, T. C. *Chem. Mater.* 2001, 13, 3516.
[26] Zhang, Q; Fu, Q; Jiang, L; Lei, Y. *Polym. Int.* 2000, 49, 1561.
[27] Mittal, V. *J. App. Polym. Sci.* 2008, 107, 1350.
[28] Osman, MA; Rupp, JEP; Suter, UW. *J. Mater. Chem.* 2005, 15, 1298.
[29] Osman, MA; Mittal, V; Suter, UW. *Macromol. Chem. Phys.* 2007, 208, 68.
[30] Mittal, V. *Eur. Polym. J.* 2007, 43, 3727.
[31] Mittal, V. *J. Colloid Interface Sci.* 2007, 315, 135.
[32] Akelah, A; Moet, A. *J. Mater. Sci.* 1996, 31, 3589.
[33] Fu, X; Qutubuddin, S. *Polymer* 2001, 42, 807.
[34] Wang, T; Wang, M; Zhang, Z; Ge, X; Fang, Y. *Mater. Lett.* 2007, 61, 3723.
[35] Velten, U; Shelden, RA; Caseri, WR; Suter, UW; Li, Y. *Macromolecules* 1999, 32, 3590.
[36] Velten, U; Tossati, S; Shelden, RA; Caseri, WR; Suter, UW; Hermann, R; Müller, M. *Langmuir* 1999, 15, 6940.
[37] Uthirakumar, P; Kim, CJ; Nahm, KS; Hahn, YB; Lee, YS. *Colloids Surf. A: Physicochem. Eng. Aspects* 2004, 247, 69.
[38] Mittal, V. *J. Colloid Interface Sci.* 2007, 314, 141.
[39] Fan, X; Xia, C; Fulghum, T; Park, MK; Locklin, J; Advincula, RC. *Langmuir* 2003, 19, 916.
[40] Zhao, H; Argoti, SD; Farrell, BP; Shipp, DA. *J. Polym. Sci., Part A: Polym. Chem.* 2004, 42, 916.
[41] Baottcher, H; Hallensleben, ML; Nuss, S; Wurm, H; Bauer, J; Behrens, P. *J. Mater. Chem.* 2002, 12, 1351.
[42] Georges, MK; Lukkarila, JL; Szkurhan, AR. *Macromolecules* 2004, 37, 1297.
[43] Husseman, M; Malmstrom, EE; McNamara, M; Mate, M; Mecerreyes, D; Benoit, DG; Hedrick, JL; Mansky, P; Huang, E; Russell, TP; Hawker, CJ. *Macromolecules* 1999, 32, 1424.
[44] Mittal, V; Herle, V. *J. Colloid Interface Sci.* 2008, 327, 295.

In: Polymer Nanocomposites: Advances in Filler Surface
Editor: Vikas Mittal
ISBN: 978-1-60876-125-8


*Chapter II*

# Exfoliation through Esterification for Clay Polymer Nanocomposites

***B.S. Jai Prakash**, Y.S. Bhat and C. Ravindra Reddy*[1]**
Clay Research Laboratory, Bangalore Institute of Technology,
K.R. Road, Bangalore-560 004, India
[1]Present address: Department of Chemical Engineering, University of Waterloo, Waterloo, ON, N2L3G1, Canada

## Abstract

Intense research activity in the field of nano composites has gained momentum in the recent years because of their enhanced properties compared to the pure polymers and the conventional fibre filler reinforced composites. In the early 1990s Toyota Researchers discovered a new class of materials with superior properties based on incorporation of nano-sized clay particles into nylon polymer matrix. Since then, many attempts are being made to understand the exfoliation of clay layers in order to incorporate them in the polymer matrix with the basic objective to prepare polymer-clay nanocomposites (PCNs) as advanced materials with much improved mechanical and thermal properties.

Swelling clays such as montmorillonite contain aggregated clay sheets having thickness in the nano range. Exfoliation of the aggregated clay sheets would result in clay sheets of nano thickness which when uniformly distributed in a polymer matrix would result in PCNs. Understanding the various methods to completely exfoliate the clay sheets and distributing the latter uniformly is a challenging task and requires the understanding of the basic work carried out in intercalation and interlayer chemistry involved in organo- clays. Esterification is one of the processes which are readily catalysed by clays which can be very easily modified to generate acid sites. . Preparation of esters including large ones can be carried out on the surface of clays including the interlayer region where suitable ambience could be created through different types of modification. This could be advantageously used to bring about exfoliation. The objective of this chapter is to review the work that has been carried out in the recent

* Corresponding author: jprak27@yahoo.com

years on the different types of surface modifications and other techniques and practices employed to bring about esterification processs on the clay surfaces that possibly could be employed to achieve exfoliation desired for preparation of polymer clay nanocomposites.

**Keywords**: Montmorillonite, organo-clay, exfoliation, nanocomposites, esterification.

## 2.1. Introduction

Polymer-clay nanocomposites are a new class of reinforced plastics formed by dispersion of nano-sized clay particles in a polymer matrix which have received considerable attention in both scientific and industrial areas since the first successful invention of nylon-clay nanocomposites by Toyota researchers in the early 1990s [1-4]. The main reason behind the intense research activity in this field is the enhanced properties of nanocomposites compared to the pure polymers and the conventional fibre reinforced composites. The Toyota researchers noticed in their investigation that nylon-6 clay nanocomposites displayed pronounced improvement in mechanical and thermal properties. They observed increase in the tensile strength by 40%, flexural strength by 60%, tensile modulus by 68%, flexural modulus by 126% and heat distortion temperature from 65°C to 152°C on addition of just 5% clay to that of pure nylon-6 matrix. The enhanced properties of polymer-clay nanocomposites are due to high interfacial interaction between the homogeneously dispersed nano-size clay platelets and polymer matrix. In addition to good thermal and mechanical properties, polymer-clay nanocomposites display good dimensional stability, barrier properties for gases and liquids, heat distortion temperature, flame retardancy and reduced smoke emission, chemical resistance and optical clarity compared to conventional composites [5, 6].

For compatibility reasons, the clay needs to be modified suitably with organic species in order to use in the preparation of clay based polymer nanocomposites. The surface modification of clay is one of the most important steps in the preparation of nanocomposites. Hence, the surface modification of clays has received considerable attention of materials scientists because it allows the creation of new materials and new applications in the field of polymer-clay nanocomposites. Further, modified clays are also used in other applications such as adsorbents for removal of organic pollutants from soil, water and air, rheological control agents, paints, cosmetics, drug delivery systems, refractory varnish, and thixotropic fluids [7]. Several modifiers and routes have been employed in the preparation of modified clays to meet the requirements of specific application. This section chapter reviews the general classes of modifiers used in the preparation of organo-clays emphasizing the utilization of esterification in preparation of clay based polymer nanocomposites.

### 2.1.1. Clay Structure and General Properties

Montmorillonite is the most widely used clay in the preparation of polymer-clay nanocomposites. Montmorillonite belongs to smectite group, which has 2:1 layered structure. Each layer constitutes one platelet and it consists of an octahedral brucite/gibbsite sheet fused between two tetrahedral silicate sheets. As a consequence of isomorphous substitution of low valent atoms like $Mg^{2+}$ for $Al^{3+}$ in the octahedral sites and $Al^{3+}$ for $Si^{4+}$ in the tetrahedral sites, the layers acquire a fixed negative charge, which is compensated by counter cations such as $Na^{+}$, $K^{+}$ or $Ca^{2+}$[8]. The counter cations occupy the interlamellar region in their hydrated form and which are exchangeable with other inorganic and organic cations. The structure of montmorillonite is presented in Figure 1 [9]. In general, the natural montmorillonite consists of aggregates of about 100 microns in diameter. Aggregates are made up of several primary particles of 1-10 microns in size. Further, each primary particle is composed of tactoids of 0.1-1 microns, which in turn consists of few tens of platelets. The clay platelets are held by electrostatic and van der Waals forces and which are relatively weak in montmorillonite and hence the interlayer distance varies depending on the radius of the cation present and its degree of hydration.

The thickness of the platelets is a well defined crystallographic dimension while the lateral dimensions of the platelet are not. Many authors in the literature have grossly exaggerated lateral dimensions of the platelets in the order of microns to few tens of microns. A commonly used montmorillonite is characterized completely for lateral dimensions by using atomic force microscopy [10]. The platelets of clay are not uniform or regular in size and hence the distribution of length is found. Most of the platelets have length in the order of 100-250 nm and the rest are in between 250-500 nm. Hence the clay platelets have high aspect ratio in the order of 100-500 nm, which is one of the most desired requirements for reinforcement. Clay platelets also have high surface area in the range of 700-800 $m^2/g$. High aspect ratio and surface area of clay platelet are primary requirements for better interfacial interaction between clay platelet and polymer matrix. The clay platelets have very good strength and stiffness along the layer direction [8]. Further, the individual clay particles themselves are regarded as very rigid inorganic polymers with very high molecular weight in the order of $1.3 \times 10^8$ which is greater than typical polymers. In addition to these properties, clays have rich intercalation chemistry which allows them to modify organically to make compatible with polymers.

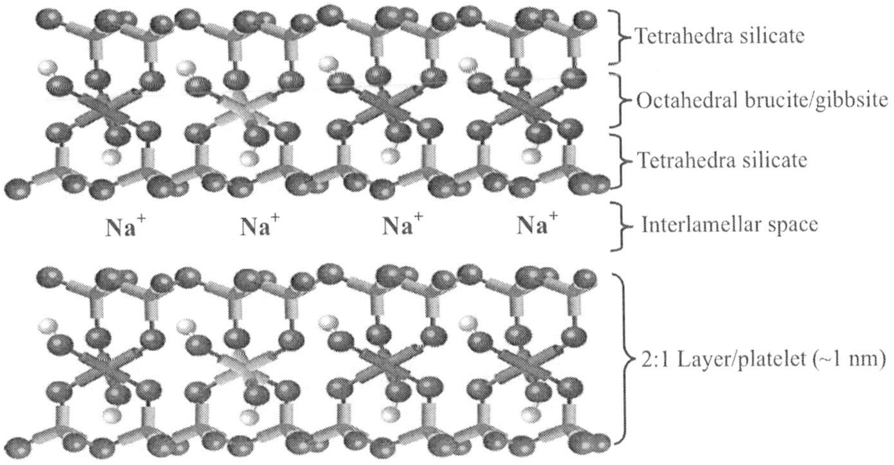


Figure 1. Reprinted from structure of montmorillonite, Polymer, 49, D.R. Paul and L.M. Robeson, Polymer Nanotechnology: Nanocomposites, 3187-3204 (2008) with permission from Elsevier.

### 2.1.2. Processing of Polymer-Clay Nanocomposites

Polymer-clay nanocomposites are currently processed mainly by three techniques: melt blending, solution induced intercalation and *in in-situ* polymerization [11]. Among three processing techniques, melt blending is the most extensively studied one. In this technique, the polymer matrix gets intercalated into the clay galleries, when organo-clay is mixed with the molten polymer matrix. The commercial polymer processing industry can apply this technique to produce polymer-clay nanocomposites based on conventional polymer processing techniques such as extrusion and injection moulding. The solution intercalation is less extensively studied technique. In this method, first, the organo-clay is swollen using suitable solvent. Later on, the solution of polymer in the same solvent is prepared and mixed with the organo-clay suspension in order to intercalate polymer into the clay galleries to get polymer-clay nanocomposite. This technique poses difficulties for commercial production of polymer-clay nanocomposites, because the solvents used are expensive and the process needs the isolation of the final polymer-clay nanocomposite from the solution. However, solution intercalation method is more useful for water soluble polymers because, water is less expensive solvent with minimum or negligible health and safety risks. *In-situ* polymerization is a useful technique to obtain polymer-clay nanocomposites. The method involves the incorporation of polymer precursor between clay layers and then expanding and dispersing the clay platelets in to the polymer matrix by polymerization. The advantage of this technique is, it produces well exfoliated structure and has been applied to wide range of polymer systems such as nylon-6, epoxy, polyurethane, polystyrene and polyethylene terephthalate. The *In iIn-situ* polymerization is more useful for obtaining clay nanocomposites of thermosetting and insoluble polymers as the method is expected to produce more exfoliated composite. Direct melt blending and *in-situ* polymerization are most important polymer-clay nanocomposites processing techniques because of their low cost, high productivity and compatibility with current polymer processing techniques.

In addition to these processing techniques, newer techniques such as sol-gel method [12-14], solid intercalation [15] and co-vulcanization [16] are also known.

### 2.1.3. Morphology of Polymer-Clay Nanocomposites

When clays are dispersed into a polymer matrix, two kinds of composites, either conventional or nanocomposite can be formed depending on the nature of components of composite and process conditions. Conventional composite is obtained if polymer is not intercalated into a clay interlamellar region (Figure 2). In such conventional composites, the clay particles act as fillers and get dispersed in the polymer matrix. The fillers being micro structured get dispersed in polymer matrix and the properties of resulting composite are similar to that of conventional composite obtained by reinforcing any micro particles. The nanocomposite is obtained when a polymer intercalates in the interlamellar region of clays or clay platelets; the latter separates and gets dispersed in polymer matrix. Thus two types of nanostructures are possible during the synthesis of polymer-clay nanocomposites - intercalated and exfoliated.

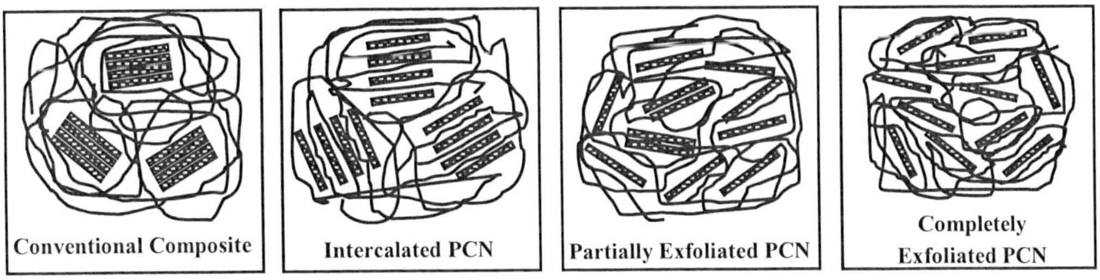


Figure 2. Morphology of polymer clay conventional composite and nanocomposites.

In the intercalated structure, single layer of polymer chain is inserted into the gallery of clay mineral resulting in a multilayer morphology with polymer layers and clay platelets stacked alternately at a repeating distance of few nanometres. In this intercalated form, there is will be an expansion of clay sheets but still there exists some interaction between clay sheets. In case of exfoliated structure, the clay sheets get separated completely into individual platelets and dispersed uniformly in a continuous polymer matrix. However, there is an intermediate nanostructure in which the clay is partially exfoliated. This is common in most polymer clay nanocomposites. The partially exfoliated nanostructure is sometimes referred to as cluster structure. Generally, the morphology of polymer-clay nanocomposites is characterized by wide angle X-ray diffraction (WAXRD) and transmission electron microscopy (TEM) [17-19]. Position, shape and intensity of basal reflections resulting from clay platelets are analyzed by the XRD pattern and the structure of composites such as conventional, intercalated and exfoliated can be identified. On the other hand TEM analysis of composites helps in the understanding of internal structure and spatial distribution of clay platelets in the polymer matrix. The typical WAXRD patterns and corresponding TEM images of different types of composites are reproduced in Figure 3.

### 2.1.4. Organo-Clays

Modification of clay surface with organic compounds is one of the crucial steps in preparing clay based polymer nanocomposites. Since they acquire organophilic nature they are also referred to as organo-clays. Natural clays are hydrophilic in nature due to presence of silanol groups on clay surface and hydrated cations at interlamellar space. The polymer-clay nanocomposites of water soluble polymers such as polyethylene oxide [20, 21] and polyvinyl alcohol [22, 23] can be obtained directly from natural clays without any modification. However, in most of the polymer-clay nanocomposites processing, modification of clay is necessary in order to convert hydrophilic clays into organophilic and to make compatible with polymer matrix. The most common method of modification is treatment of natural clay with organic species generally consisting of cations including primary, secondary, tertiary, and quaternary alkylammonium cations. The modification in most of the cases is achieved by the cation exchange technique. The organic species are sometimes referred to as surfactants owing to their amphiphilic nature. These cationic surfactant species possesses long alkyl chain tail and polar head group. When clay is treated with cationic surfactant, the simple cations like $Na^{+}$ and $Ca^{2+}$ present in pristine clay will be exchanged with organic cations (Figure 4). The modified clay with surfactants acquires organophilic nature. Such a modification also results in expansion of clay platelets and decreases the interlayer energy which is an important requirement for intercalation/exfoliation of clay during polymer nanocomposite formation. The modification also improves the wetting characteristics of clay with polymer matrix. The organic species used in modification allow the thermodynamically favourable penetration of polymer into clay and assists in the delaminating of clay. Further, the modifying agents can provide functional groups that can react with the polymer matrix, or in some cases initiate the polymerization of monomers to improve the strength of the interface between the inorganic and the polymer matrix [24, 25].

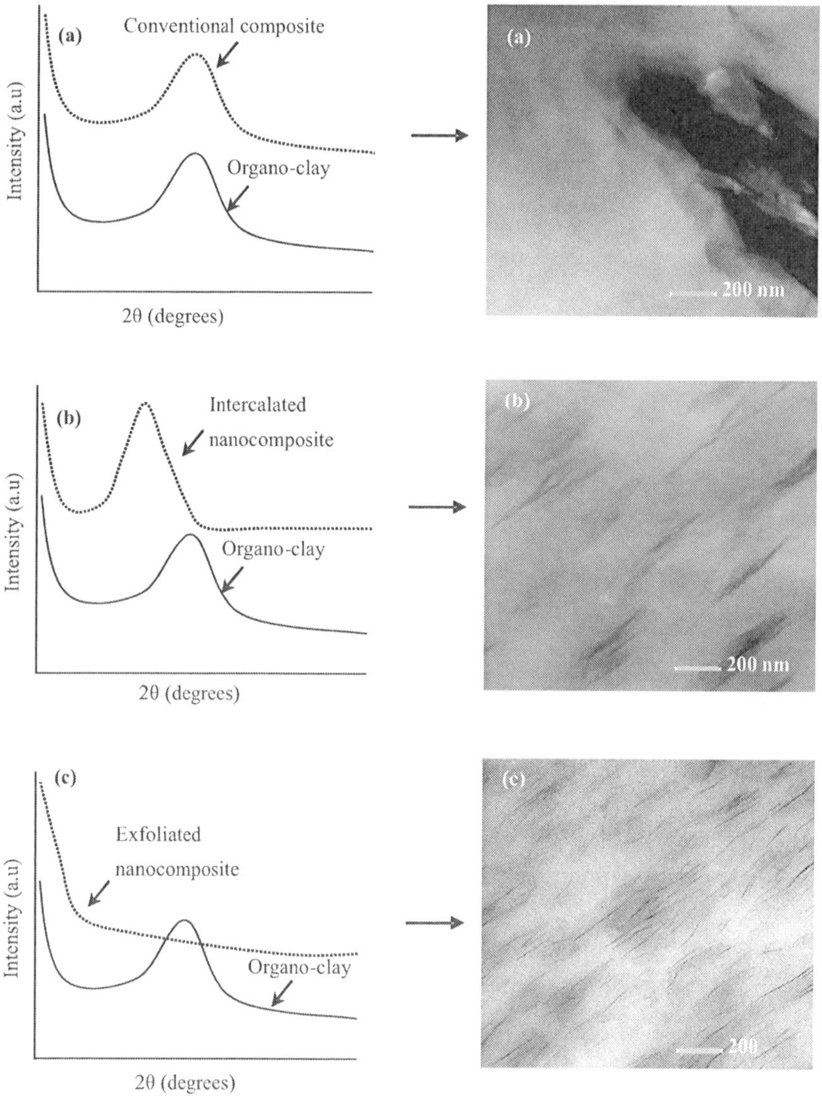


Figure 3. Reprinted from Typical XRD pattern and TEM images of (a) conventional composite, (b) Intercalated and (c) Exfoliated polymer-clay composites. Polymer, 49, D.R. Paul and L.M. Robeson, Polymer Nanotechnology: Nanocomposites, 3187-3204 (2008) with permission from Elsevier.

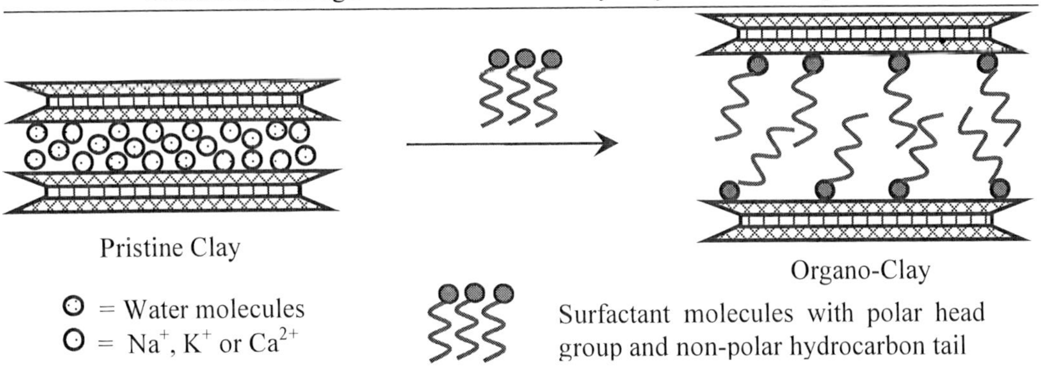


Figure 4. Schematic representation of modification of pristine clay to organo-clay.

The alkylammonium compounds are first generation organic compounds most commonly used to make clays compatible with polymer matrix [26]. The alkylammonium compounds of this class include, commercially available quaternary ammonium compounds having long alkyl chain, alkyl chain along with some functional groups such as hydroxyl, carboxylic, amine, alkoxy, ester, etc. However, these compounds possess thermal instability and hence undergo decomposition during processing of nanocomposites. In order to overcome the drawback of poor thermal stability of simple alkylammonium compounds, several other modifiers such as oligomeric alkylammonium [27-29], phosphonium [30, 31], imidazolium [32, 33], benzimidazolium [34], quinolium and pyridinium [35], stibonium [36] compounds and crown ethers [37] have been used in the preparation of organo-clays. Organo-silane grafting is also practiced in the modification of clays to organophilic in order to study them in polymer-clay nanocomposite processing [38, 39].

Modification of clays with cationic surfactants carrying functional groups and interacting the latter to link with other organic molecules in order to increase the interlamellar space further is being investigated in the preparation of polymer nanocomposites. Recently, it is suggested that high density brushes with long alkyl chains chemically bound to the surface can be generated by exchange of ammonium ions with reactive groups on the surface and then subsequently reacting to attach further long chains [(40]).

## 2.2. Exfoliation through Esterification

### 2.2.1. Criterion to Bring about the Exfoliation

The factors influencing the exfoliation in the epoxy based clay nanocomposites is summarized by Ke and Stroeve [41]. According to them, there should be enough epoxy molecules participating in the reaction in the clay micro region system and the reaction rate inside be more rapid than outside interlayer gallery in order to discharge enough heat in the inside space, overcoming the van der Waals force between the interlayers, exfoliating the clay into monolayer and dispersing in the epoxy matrix.

The difficulty in the preparation of the polyester-clay nanocomposites is the exfoliation uniformly throughout the polymer matrix into individual layered silicates as the required energy necessary to exfoliate may not be generated during the formation of the polyester.

However, ester formation as a means of interlayer space fill and establishing an organophilic environment is suggested by many [60, 65]. The ester attached to the clay layer will enhance the interlayer distance and help to create an organophilic environment around the clay sites thus facilitating the entry of the monomers/oligomers or the polymer resin. In the case of ester formation two components are required, one a reactive carboxylic acid (–COOH) component and another reactive –OH component. The first logical step would be to introduce alkylamines with functional groups which can be easily esterified (a reactive surfactant monomer). Alkyl amines are known to readily occupy the interlayer in excess of cation exchange capacity (CEC). The interlayer is thus packed with the alkyl amine chains imparting a hydrophobic environment and pushing away the clay layers apart but not sufficient to bring about exfoliation. The second step is to manipulate the interlayer environment to fill the hydrophobic space by some means to further push away the layers and at the same time liberating enough enthalpy required to exfoliate. The hydrophobic environment can now admit the polymer resin (matrix) to space fill and effectively bring about the total exfoliation and disperse the clay platelets. One method involves the introduction of a suitable catalyst that can bring about the esterification of the functional groups of the surfactant with the reactive acid components in the medium and, as a result, liberate enough energy to overcome the resistance to exfoliation. This implies introducing an acid catalyst component in the interlayer prior to the reactive surfactant and then admitting the latter to create an hydrophobic environment. The two together in the interlayer will not only facilitate the entry of the other reactive component but also favours the esterification. Thus, a favourable organophilic environment is created for the entry of the monomer unit of the polymer resin to space fill the interlayer and to complete the polymerization process.

Another novel way suggested is to employ surfactants with polyhydroxy groups in the interlayer and then esterify them using organophilic catalysts. It has been observed that increasing the number of hydroxyl groups in combination with the hydrocarbon chain present in the modification was successful in generating high density brushes on the clay surface which also was responsible for achieving higher basal plane spacing of the montmorillonite platelets owing to the reduction of electrostatic interactions holding them [65]. A careful choice of the catalyst (similar to a curing agent in the case of epoxy) and the reactive ester components will thus facilitate the process. Understanding the incorporation of the different acid centres in the interlayer and the methods followed to bring about simple esterification in different clay materials is useful to bring about the preliminary step that is to develop an organophilic environment. The work carried out on esterification and polyester formation is reviewed in the ensuing sections.

### 2.2.2. Esterification on Clay Surfaces

An attempt is made here to review the work carried out by different workers on the generation of acid sites for catalytic functionality to bring about esterification on clay surfaces and thus create an organophilic environment as a preliminary step for exfoliation.

In general, esterification is an acid catalyzed reaction. In this context the clay that has to act as catalyst should possess acid centers. Several publications dealing with use of clay as catalyst in esterification have been reported in which clay has been appropriately modified to acquire the acidic function. The methods used can be divided into two types, one in which clay provides its surface for supporting acidic materials like heteropoly acids, *para*-toluene sulphonic acid etc. and another in which cationic centers are appropriately modified to generate acidic centers on clay surface and interlayers. The former technique is ideal if all the interlayer space is available for supporting the acidic material. This is important basically for utilizing the interlayer space to incorporate polymeric species and exfoliate the clay layers to have nano-level dispersion in the polymer matrix to tune the certain improvement in structural and textural properties.

In the technique that involves transformation of clay cations centers to acidic sites, two procedures are followed - first one is exchanging cations of natural clay with different multivalent cations, e.g. replacing $Na^+$ with $Ni^{2+}$, $Al^{3+}$ or $Ti^{4+}$ . These metallic cations have hydration sphere around them and have the property to dissociate the coordinated water in the hydration sphere. This results in creation of Brønsted acid centers which can catalyze the esterification reaction [42]. In the second procedure cations of natural clay are replaced with protons on treating with strong protonic acids [43]. During this acid treatment, alkali cations are replaced by protons resulting in the removal of some of the structural aluminum atoms in the clay. Enough care is necessary not to severely damage the clay structural integrity.

The mechanism of clay catalyzed esterification is similar to the one catalyzed by conventional protonic acids. First step is protonation of reactant acid by using proton donated by acidic center of clay followed by nucleophilic attack of hydroxyl group of the reactant alcohol.

The literature reveals that montmorillonites show expansion of interlayer with a wide variety of polar organic compounds. Extent of swelling depends on the size and charge of the exchanged cationic species and the polar liquid. [44-45]. Esterification may be easily facilitated by the alcohol reactant enhancing the interlayer distance of the clay and enabling access to interlayer cationic sites. That is why highest activity of ester is obtained when mole ratio of acid to alcohol in esterification is more than 1:1, in some cases, as part of the alcohol might be used up for swelling of the clay. A few papers illustrating the above mentioned techniques involved are discussed below.

Gregory et al. [46] have found that Wyoming bentonites exchanged with cations of high charge density are efficient and selective catalysts for the production of ethylacetate from ethylene and acetic acid.

James et al. [47] had made an early attempt to employ clay catalysts containing different cations such as aluminium, copper or ammonium sulphate, dilute sulphuric acid or chromic chloride solutions. Reactions with gaseous alkenes were carried out in a 500 $cm^3$ rocking high pressure autoclave. The carboxylic acid and catalyst were placed in the reactor, which was sealed, charged with desired pressure of alkene and heated to required temperature. Esterification is in competition with alkene oligomerization as shown in scheme 1. The number of ester products which may be obtained in a given reaction is determined by the number of stable carbocation intermediates which each alkene can produce.

Using incipient wetness technique, clay supported $InCl_3$, $GaCl_3$, $FeCl_3$ and $ZnCl_2$ were prepared by impregnating respective metal chloride from its acetonitrile solution on montmorillonite and kaolinite by Choudhary Vasant et al., [48]. Esterification of *tert*-butanol by acetic anhydride over different catalysts was carried out in a magnetically stirred glass reactor fitted with reflux condenser and thermometer. All the clay supported metal chloride catalysts showed high selectivity but differs in their esterification activity. Among the catalysts, $InCl_3$/Mont-K-10 showed the best performance. The process is cleaner and requires much less amount of catalyst and time than that reported earlier.

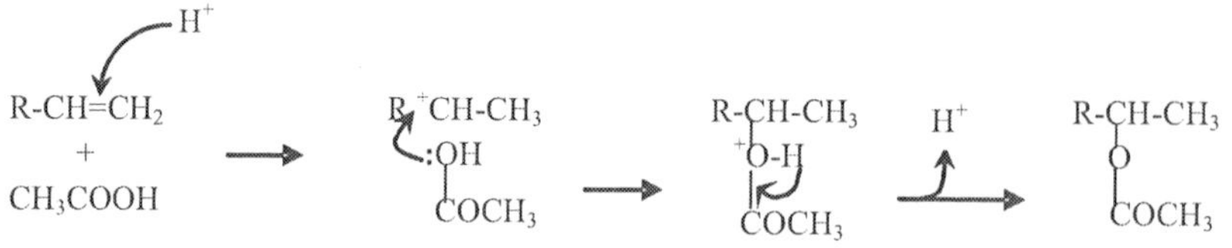


Scheme 1. Esterification of alkene with carboxylic acid [47].

Yadav and Joshi [49] have used a variety of solid acid catalysts such as K-10-montmorillonite, dodecatungstophosphoric acid supported on K-10, sulphated zirconia and ion-exchange resins were used for the acylation of resorcinol with acetic acid. To establish intrinsic kinetics of reaction the effects of various parameters on reaction rate and selectivity were investigated. The rate constant and energy of activation for the esterification have been deduced. Among the different catalysts used ion-exchange resin was found to be most effective.

Srinivas and Biswanath Das [50] have reported that $Fe^{3+}$-K10-montmorillonite as a highly efficient catalyst for esterification of aliphatic carboxylic acids in the presence of aromatic carboxylic acids. They have studied K-10-montmorillonite obtained from Fluka Chemicals as catalyst. To a solution of 2-carboxyphenyl propanoic acid (1 mmol, 194 mg) in ethanol (10 ml) was added 100 mg of clay and the mixture was refluxed for 8 h. Different carboxylic acids, both conjugated and nonconjugated were treated with various alcohols in presence of clay catalyst to obtain the corresponding esters in high yields.

Polyester-clay nanocomposites were reported to be prepared by dispersing methyl tallow bis-2-hydroxyethyl quarternary ammonium chloride modified montmorillonite in prepromoted polyester resin and subsequently crosslinked at room temperature [51]. The formation of a nanocomposite was confirmed by x-ray diffraction. The TEM of the polyester nanocomposite containing 10% clay, as obtained by the authors, is presented in Figure 5.

Seok-Ho Hwang et al. [52] have carried out synthesis of nanocomposites by *in-situ* tethering allylester to layered silicate. Organically modified layered silicate utilized in the study was bis(2-hydroxy-ethyl) methyl tallow ammonium montmorillonite (Cloisite 30B from Southern clay products). Tethering of allylester involved pouring of diallyl terephthalate and 1,3 butanediol into a three-neck round bottom flask equipped with a dean-stark trap followed by gradual addition of organo-clay. The mixture in the round bottom flask was stirred at 60°C for 4 h. In order to effect the *in-situ* polymerization catalytic material monobutyl tin oxide was added and temperature of the mixture was raised to190°C under nitrogen atmosphere. After maintaining the temperature for 5 h, the unreacted diol and remaining allylalcohol were removed under vacuum. The reaction scheme as suggested by the authors is presented in Scheme 2.

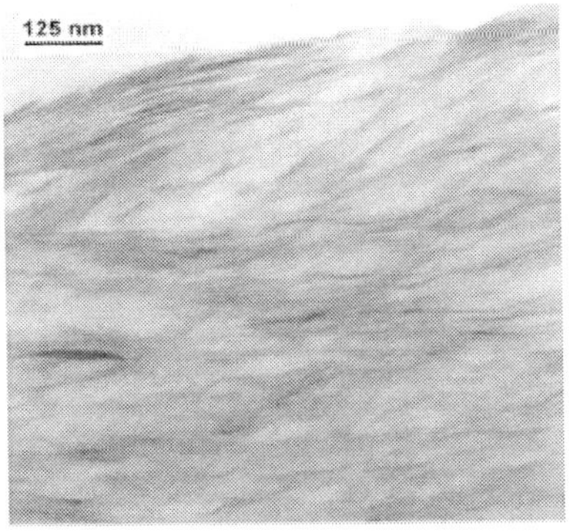


Figure 5. Reprinted from TEM micrograph of cross linked polyester nanocomposite containing 10% clay. Polymer, 43 , R.K. Bharadwaj, A.R. Mehrabi, C. Hamilton, C. Trujillo, M. Murga, R. Fan, A. Chavira and A.K. Thompson. Structure-property relationships in cross-linked polyester-clay nanocomposites, 43, 3699-3705 (2002) with permission from Elsevier.

Wang and Li [53], have used $Al^{3+}$ exchanged pillared clay catalyst for esterification of 2-ethoxyethanol with acetic acid. Chinese bentonite was used to prepare activated clay. Further it was used for the preparation of $Al^{3+}$ exchanged pillared clay by treating with aluminum sulphate of known concentration at 100°C temperature, followed by thermal treatment in the temperature range 100-400°C. For esterification a three neck round bottom flask equipped with thermometer, mechanical stirrer and water segregator was used. In a typical experiment 0.5 mol of 2-ethoxyethanol was mixed with 0.6 mol acetic acid, 3.76 g Al-pillared catalyst and 0.18 mol benzene. The catalytic activity is due to Bronsted acid sites of surface hydroxyl groups of pillared clay and Lewis acid sites due to pillaring.

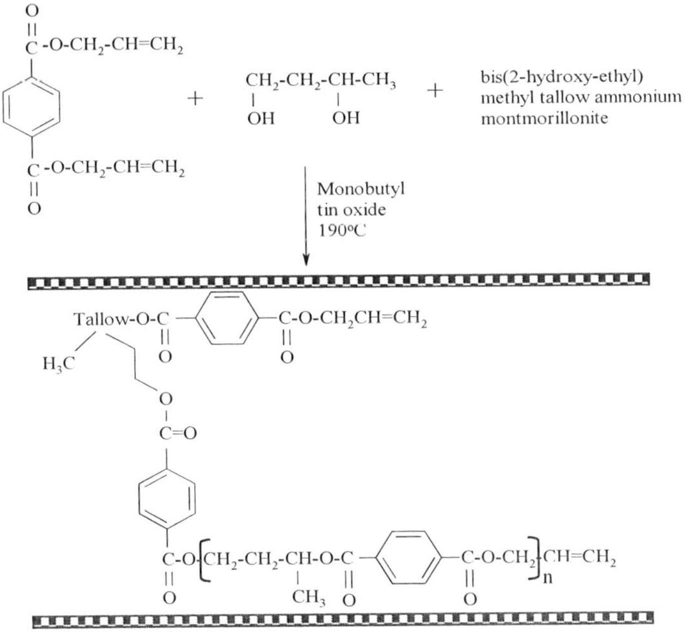


Scheme 2. Reaction scheme for *in-situ* formation of polyester-clay nanocomposites [52].

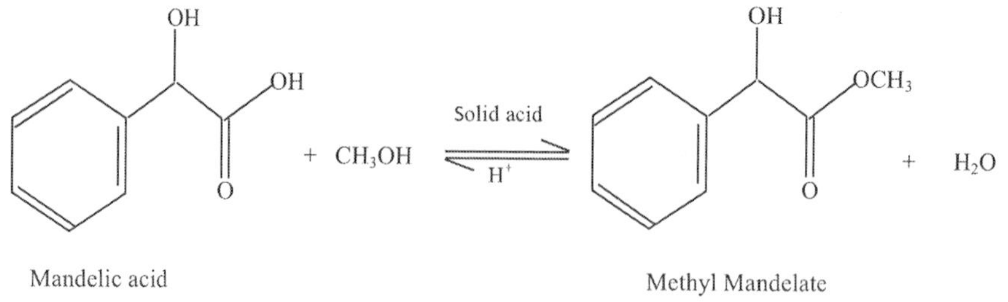


Scheme 3. Esterification of mandelic acid with methanol over solid acid catalysts [56].

Esterification of phenylacetic acid with phenol, cresols, nitrophenol and resorcinol has been carried out over $Al^{3+}$-montmorillonite and other montmorillonite modified catalytic systems [54]. $Na^{+}$-montmorillonite was inactive, $H^{+}$ and $Al^{3+}$-montmorillonite dried at 100°C, showed 52 and 67% conversion to ester, respectively. While $Al^{3+}$-montmorillonite dried at 200°C showed a conversion of 36%, the same catalyst dried at 400°C did not show any conversion. Among the cresols, the *para* isomer showed the highest yield whereas the *ortho* isomer exhibited negligible conversion possibly due to steric factors. Nitrophenol did not give ester may be due to nitro group which has the negative effect on the rate of reaction because its electron withdrawing property reduces the nucleophilic character of phenol. In case of resorcinol protonation makes it less nucleophilic.

Vijayakumar et al., [55] using an acid activated Indian bentonite, synthesized *p*-tolyl esters of some long chain fatty acid having commercial value. The effect of reactant molar ratio and reaction time were investigated. The spent catalyst on regeneration showed the same activity.

A process for synthesis of methyl mandelate was developed by using solid acid $Cs_{2.5}H_{0.5}PW_{12}O_{40}$ supported on clay as catalyst Ganapati et al. [56] for esterification of mandelic acid with methanol [56]. The effects of various parameters were studied to deduce the intrinsic kinetics.

A 0.05 mole mandelic acid, 1.5 mole methanol and 2.17 g catalyst were charged to 100 $cm^3$ Parr autoclave and heated to 120°C (Scheme 3). The reaction was carried out with stirring at 600 rpm for 5 h. The reaction mixture was analysed by HPLC. More than 80% conversion is reported for a reaction period of three hours at 0.03 g $cm^{-3}$ catalyst loading.

Ken et al. [57] have studied the combination of acid layered clay with a basic layered clay for one-pot sequential reactions for synthesising epoxynitrile. The objective of the work was to design heterogeneous catalyst by focusing on layered clays, which can generate strong acid or base sites for catalytic functionality. A combination of $Ti^{4+}$ montmorillonite and hydrotalcite (HT) as acid base catalyst was used in one pot synthesis. The $Ti^{4+}$ exchanged montmorillonite is an excellent solid Brønsted acid catalyst as a result of $Ti^{4+}$ cations in the interlayer spaces. Hydrotalcite is a good surface tunable solid basic system. Using this catalyst an intermediate for various heterocyclics, epoxynitrile was synthesized using cyanoacetic acid, methanol, aldol and hydrogen peroxide in four sequential acid base reactions. The scheme of the reactions is given in Scheme 4.

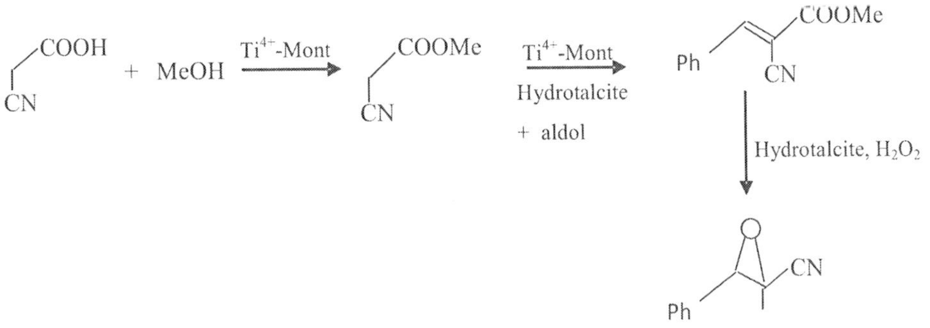


Scheme 4. Reactions scheme for the synthesis of epoxynitrile [57].

Mutual neutralization of acid and base catalysts can be avoided in a single reaction system using this unique catalyst system. Isolated catalytically active centers could be created based on clay materials for realization of acid-base tandem reactions.

Dodecatungstophosphoric acid supported on clay was employed for esterification of *p*-cresol with phenylacetic acid to form *p*-cresylphenyl acetate and its performance was compared with sulphated zirconia, ion-exchange resins Amberlyst-15 and Indion-125 [58]. The reaction is shown in Scheme 5.

Ravindra Reddy et al. [59], carried out the esterification of succinic anhydride to di-(*p*-cresyl) succinate over $M^{n+}$-montmorillonite catalysts. The activity of the $M^{n+}$-montmorillonite catalysts was found to be linearly related to the charge to radius ratio of $M^{n+}$ ions (Figure 6). $Al^{3+}$-montmorillonite is the most effective catalyst for the esterification with a maximum yield of 75%, which is comparable to the yield from the reaction conducted with *p*-toluenesulphonic acid.

Tsung-yen et al. [60] have observed that any slight step up in the dispersion of nano clay platelets in the polymer matrix would reflect in the improvement in the desirable properties over their micro and macrocomposite counterparts. They are of the opinion that seldom have studies been reported on the preparation and characterization of exfoliated poly(ethylene terephthalate) nanocomposites. The bottleneck in the preparation of these nanocomposites is how to exfoliate the clay uniformly throughout the polymer matrix into individually layered silicates. In the traditional methods, long-chain surfactants not only increase the interlayer distance but also provide compatibility for the polymerization to occur.

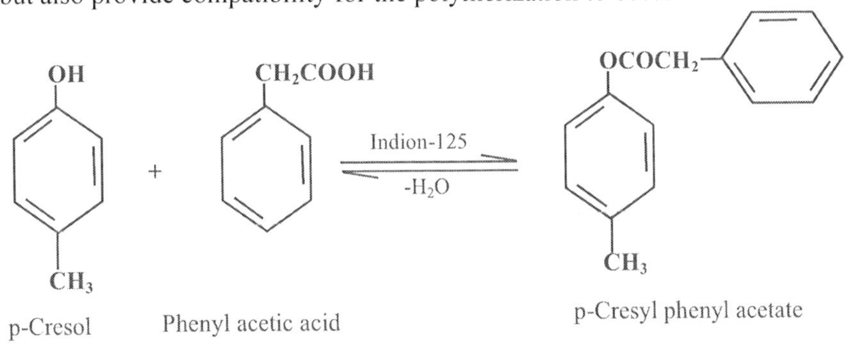


Scheme 5. Esterification of phenylacetic acid with *p*-cresol over solid acid catalysts [58].

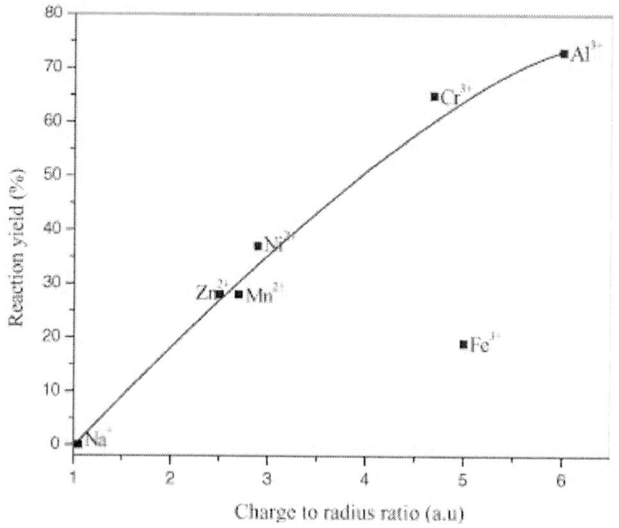


Figure 6. Reprinted from reaction yield on the esterification of succinic anhydride to di-(*p*-cresyl) succinate over $M^{n+}$-montmorillonite catalysts as a function of charge to radius ratio of the exchangeable cations. J. Mol. Catal. A: Chemical 229, C. R.avindra Reddy, P. ushpa Iyengar, G. opalpur Nagendrappa and B.S. Jai Prakash, Esterification of succinic anhydride to di-(p-cresyl) succinate over $M^{n+}$ - montmorillonite clay catalysts, 31-37 (2005), with permission from Elsevier.

According to Tsai et al.,(60) traditional routes may not be advantageous for polyester-clay nanocomposites because there is no driving force provided that can facilitate absorption of the monomers or oligomers between adjacent silicate layers during the polymerization process. The key factor concerning the diffusion of the monomer into the interlayer for polymerization is overlooked. Thus the conventional treatment causes the polyethylene terephthalate, PET/clay nanocomposites to become only an intercalate dispersion of clay instead of an exfoliated dispersion of the clay in the polymer substrate. To overcome this, Tsai et al., have intercalated a catalyst for a novel polymerization process within the gallery to proportionally disperse the stratiform clay platelets in the polyester and thus achieve exfoliation. This driving force concept makes use of a catalyst precursor such as antimony acetate $Sb(OAC)_3$. . Considerable swelling was noticed when the $Sb(OAC)_3$-treated clays were melt mixed with terphthalic acid bis(2-hydroxyethylester) [BHET] and a surfactant, acidified sodium cocoamphohydroxypropylsulfonate (SB). The exfoliation was confirmed by XRD patterns where the (001) reflection at $2\theta = 6.1^0$ corresponding to d spacing approximately 14.48Å for the clay composite without $Sb(OAC)_3$ that is, [PET/clay/ SB] which disappears for the clay composite with $Sb(OAC)_3$ that is, [PET/clay/Sb/ SB] showing exfoliation in the latter.

This is further confirmed by the transmission electron micrographs of the samples. Fig.8 illustrates the proposed mechanism for the formation of polymer-clay nanocomposites by the driving force concept by Tsai et al [60].

Tao et al. [61], have shown that acidified clay displayed a higher activity for esterification of cyclo-2-pentene with acetic acid. The mixture of cyclo-2-pentene 0.7 ml (10 mmol) and acetic acid 1.75 ml (30 mmol) was stirred at room temperature in 25 ml flask. To this clay was added and reaction was carried out at 80°C for 4 h (Scheme 6). The progress of the reaction was monitored using GCMS.

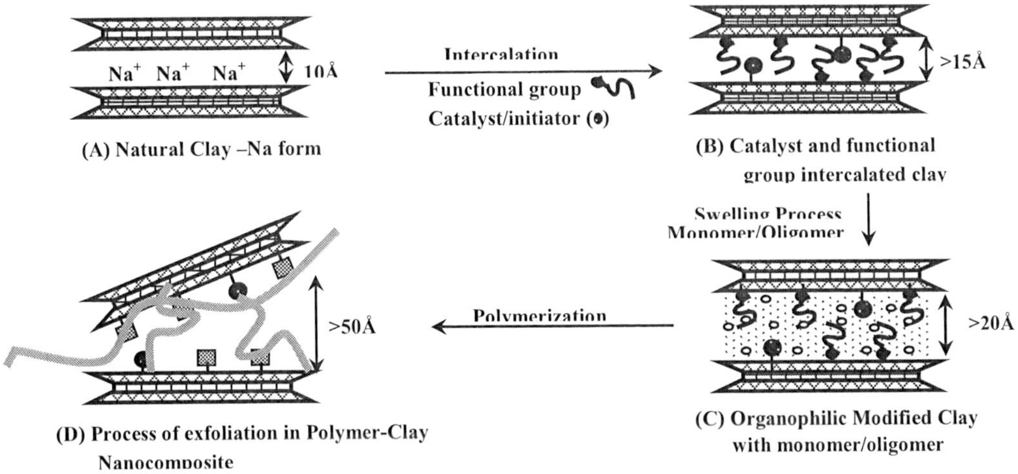


Figure 7. Illustration of driving force concept mechanism proposed by Tsai et al. for the formation of polymer /clay nanocomposites (A) Sodium-exchanged smectite clay. (B) Interlayer expansion due to intercalation of catalyst/initiator and modified agents with functional groups (C) Creation of organophilic interlayer environment; monomers/oligomers driven by the catalyst/initiator to swell into the gallery of the clay lamellar. (D) Exfoliation of clay platelets owing to polymerization occurring between the adjacent silicate layers in the polymer matrix.

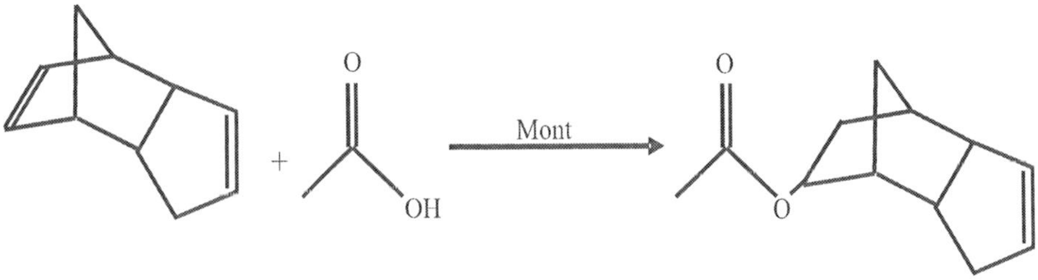


Scheme 6. Esterification of cyclo-2-pentene with acetic acid over montmorillonite [61].

Due to possession of stronger acid sites, acid treated clay displayed a better esterification activity. Theoretical analysis showed that tetrahedral Si plays an important role in supporting the framework for montmorillonite, and the process of substitution will occur apart from octahedral aluminium.

One of the ways to prepare clay nanocomposites is exfoliation through polymerization like esterification of diols and diacids. This approach typically involves dispersing the nanofiller, usually a smectite like a montmorillonite, in one or more of the monomers and subsequently forming the polymer around the dispersion. One of the keys to successfully exfoliating the clay with this process involves selecting the proper intercalating agent. The interaction between the intercalating agent and the monomer must be sufficiently strong so that it is capable of driving the monomer into the galleries of the clay. This was followed for barrier applications by David et al. [62].

A stainless steel autoclave was charged with dimethylterephthalate, DMT (10.1 lbs, 4.59 kg), ethylene glycol (6.7 lbs, 3.0 kg), antimony trioxide (2.80 g), manganese acetate (3.60 g), sodium acetate (1.30 g), and PANGEL® B20 sepiolite (140.0 g). The reaction vessel was purged with 60 psi of nitrogen three times. The vessel was heated to 240°C with a low flow nitrogen sweep of the vessel. While the vessel was heating to 240°C, the reaction was agitated at 25 RPM. After the vessel reached 240°C, the reaction temperature was maintained for 10 min. The reaction mixture was then heated to 275°C and a 90 minute vacuum reduction cycle was begun. Upon completion of the vacuum reduction cycle, a full vacuum (0.1 torr) was applied to the reaction and the reaction mixture was maintained at 275°C for 120 min. The reaction was pressurized with nitrogen and the polymer was extruded as a strand, cooled in a water trough, and chopped into pellet form. In this process DMT is source for diacid terephthalic acid, ethylene glycol is the diol and antimony oxide and metal acetates are the catalysts for esterification.

Philip et al. [63], using K-10 batches showed varying activity in Brønsted acid catalysed reactions due to variation in acid treatment conditions. It is subjected again to acid treatment to enhance its catalytic activity and tested for three reactions viz. tetrahydropyranylation of ethanol, diacetylation of benzaldehyde and esterification of succinic anhydride. Acid treatment of clay resulted in protonation and loss of layer stacking of the clay structure.

Esterification was effected by adding known quantity of acid treated clay to succinic anhydride and *p*-cresol dissolved in toluene and refluxing the solution. Two pellets of clay are also added before reflux to adsorb water formed during the reaction. Esterification generally increased with strength of acid treatment, indicating the importance of highly acid treated material for this reaction.

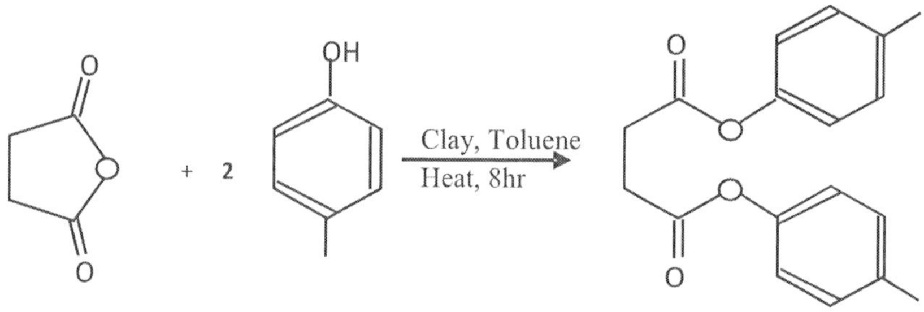


Scheme 7. Esterification of succinic anhydride with *p*-cresol over clay catalysts [63].

Harman et al. [64] have prepared new esters of 5-hydroxymethylfurfural with maleic acid and phthalic acid and its isomers. The scheme for this reaction is presented in Scheme 8. 5-Hydroxymethylfurfural was charged to a 250 ml round bottom flask in a glove bag under argon. With continuous flow of argon, maleic anhydride and the catalyst were added to the flask. The reaction mixture was heated to 100°C with stirring. This condition was kept for 20 h. The reaction mixture turned viscous blackish brown during this period which was analysed by proton NMR.

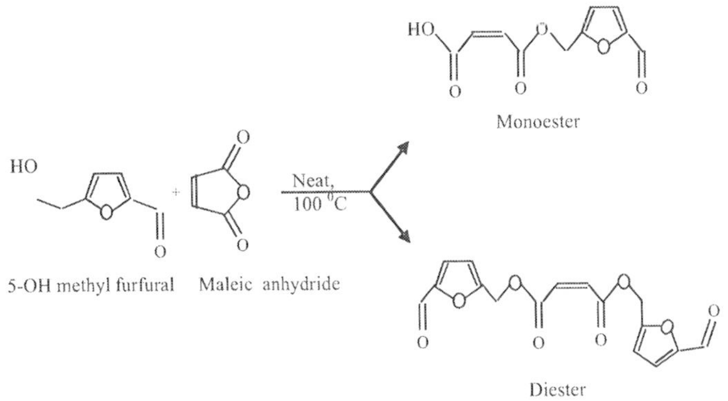


Scheme 8. Esterification of 5-Hydroxymethylfurfural with maleic anhydride [64].

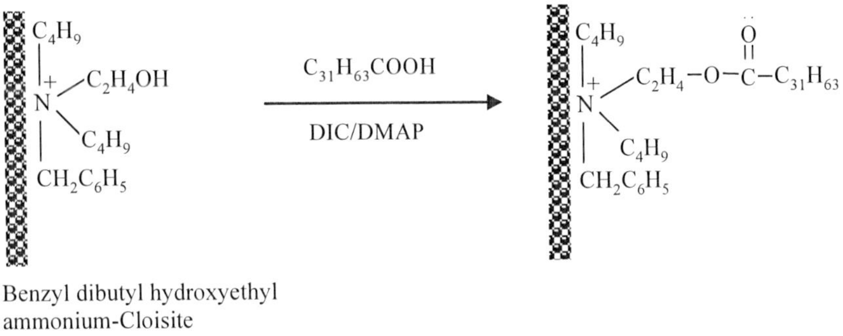


Scheme 9. Esterification of hydroxyl group of alkylammonium modifier with fatty acid [65].

In addition to using the cationic centers as acidic sites for esterification, the same reaction can be effected by intercalating large organo-cations possessing hydroxyl groups in the alkyl chain in place of smaller ones followed by esterification of organo-cation with long chain fatty acids in presence of suitable catalysts. This approach finds application in the preparation of nano composite materials containing exfoliated clay. Mittal [65] has modified the montmorillonite clay platelets with ammonium ions containing terminal hydroxyl group. This group was further esterified via surface reaction with long chain fatty acid. The enhancement in interlayer distance due to the modification was characterized by thermogravimetric analysis, X-ray diffraction and infrared spectroscopy. The objective of the work was to increase the basal spacings in the clay so as to decrease electrostatic interactions that will help the clay exfoliation in the polymer matrix.

Before the esterification organic cation was attached to the clay surface. To carry out esterification, 0.5 g of modified clay was suspended in 5 ml of dichloromethane or DMF and sonicated. To this fatty acid octanoic acid or dotriacontanoic acid was added followed by dimethyl amino pyridine and diisopropyl carboimide. The contents were kept under stirring for 24 h. The resulting thick gel was separated and treated with ether followed by DMF. The whole process was repeated to enhance the extent esterification.

The initial incorporation of ammonium compounds resulted in the swelling of the clays that helped to further modify the surface by esterification. The method helped to expand the interlamellar distance varying from 16.6 to 52.7Å. This is a novel technique to enhance interlayer spacing in the clay that is to be exfoliated within polymer matrix.

Yadav et al. [66] have studied the esterification of benzoic acid with phenol followed by Fries rearrangement leading to 4-hydroxybenzophenone (Scheme 10) in one pot synthesis under solvent free condition was carried out over cesium substituted heteropoly acid supported on clay catalyst. Benzophenone and its derivatives are useful for ultraviolet light protection in cosmetic ingredients. Commercially available K-10 clay (Fluka, Germany) was used. The clay was dried in an oven at 120°C for 3 h before incorporation of heteropoly acid. The reaction was carried out in 100 $cm^3$ stainless steel Parr autoclave having stirring facility. Effect of reaction parameters such as stirring speed, catalyst amount, temperature, mole ratio on the yield of 4-hydroxybenzophenone was studied. At a mole ratio phenol: benzoic acid 7:1 and using 0.05 g/cm3 catalyst, 70% acid got converted with selectivity of 32.5% for the desired product.

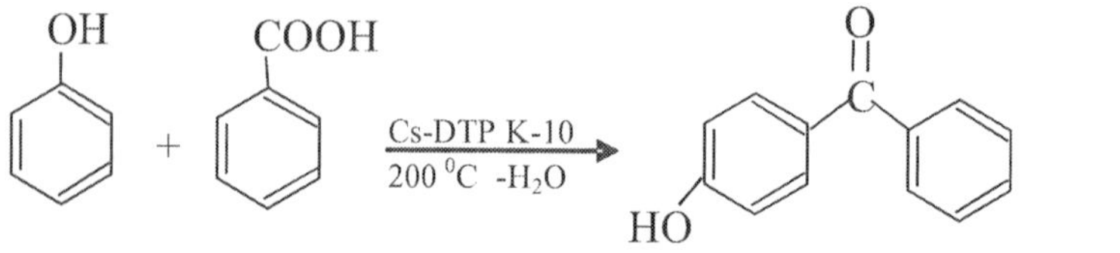


Scheme 10. Synthesis of dihydrocoumarin derivatives over clay catalyst [66].

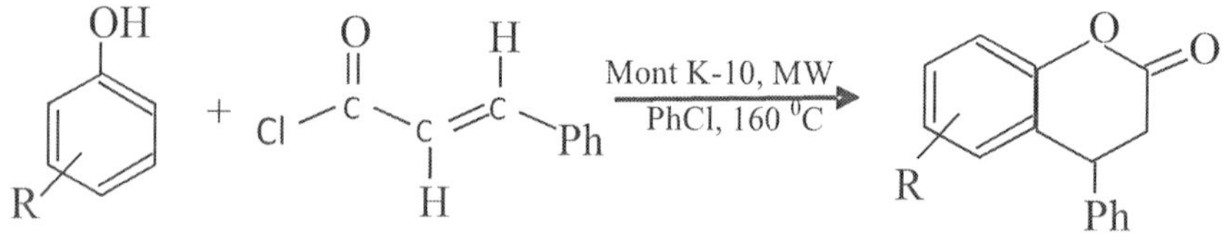


Scheme 11. Synthesis of dihydrocoumarin derivatives over clay catalyst [67].

The product selectivity was mainly dependent on duration of reaction, temperature, mole ratio and catalyst loading. Esterification is faster than Fries rearrangement and is the main pathway in a two-step route. The catalyst used was active, stable and reusable.

Zheng et al. [67] have carried out a tandem esterification-Friedel Crafts alkylation reactions (Scheme 11) under microwave irradiation over environmentally benign solid acid catalyst montmorillonite K-10 catalyst. The reactions involved are synthesis of dihydrocoumarin derivative through phenols and cinnamoyl chloride. The coumarins are important intermediates for pharmaceutical industry, useful pesticides and bioactive compounds. The reaction was effected by mixing phenol and cinnamyol chloride 2mmol each with montmorillonite K-10 in dried chlorobenzene in a sealed tube and microwave irradiated at 160°C for 5 minutes.

These reactions required long reaction time under conventional heating conditions, occurred solvent free pattern, with low reaction time and higher yield under microwave irradiation mode of heating.

## 2.3. Conclusions

In-situ intercalative polymerization is regarded as one of the best ways of preparing polymer-clay nanocomposites. Although significant changes in the desirable properties of the polymer-clay nanocomposites are reported, total exfoliation and homogeneous distribution of the clay platelets are still not achievable in all cases. Research work is in progress to modify the clay surface suitably to achieve exfoliation and uniform distribution of clay platelets throughout the polymer matrix. Modification of clay suitably with organic species carrying reactive functional groups like hydroxyl and carboxylic acid followed by esterification seems to be a feasible in preparation of exfoliated polymer-clay nanocomposites, specifically polyester-clay nanocomposites. Review of the work going on indicates that creation of a suitable environment in the interlayer to bring about esterification between the reactive monomer components to exfoliate the clay platelets is the main focus. With more work, clear cut guidelines will emerge out to choose the right combination of the ester components, catalyst, surfactant and the layered material to create the right interlayer environment. Exfoliation through esterification and subsequent uniform distribution in the polymer matrix will, consequently, be a method worth trying for making polyester-clay nanocomposites.

## References


[1] Usuki, A; Kojima, Y; Kawasumi, M; Okada, A; Fukushima, Y; Kurauchi, T. *J. Mater. Res,* 1993, 8, 1179-1184.

[2] Kojima, Y; Usuki, A; Kawasumi, M; Okada, A; Fukushima, Y; Kurauchi T; Kamigaito, O. *J. Mater. Res,* 1993, 8, 1185-1189.

[3] Kojima, Y; Usuki, A; Kawasumi, M; Okada, A; Kurauchi T; Kamigaito, O. *J. Appl. Polym, Sci,* 1993, 49, 1259-1264.

[4] Usuki, A; Kawasumi, M; Kojima Y; Okada, A. *J. Mater. Res,* 1993, 8, 1174-1178.

[5] Usuki, A; Hasegawa N; Kato, M. *Adv. Polym. Sci,* 2005, 179, 135-195.

[6] Hussain, F; Hojjati, M; Okamoto, M; Gorga, RE. *J. Comp. Mater,* 2006, 40, 1511-1575.

[7] Patel, HA; Somani, RS; Bajaj, HC; Jasra, RV. *Bull. Mater. Sci,* 2006, 29, 133-145.

[8] Pinnavaia, TJ. *Science,* 1983, 220, 365-371.

[9] Paul DR; Robeson, LM. *Polymer* 2008, 49, 3187-3204..

[10] Ploehn HJ; Liu, C. *Ind. Eng. Chem, Res,* 2006, 45, 7025-7034.

[11] Alexandre M; Dubois, P. *Mater. Sci. Eng,* 2000, 28, 1-63.

[12] Alam, SM. M; Agag, T; Kawauchi T; Takeichi, T. *React. Funct. Polym,* 2007, 67, 1218-1224.

[13] Bandyopadhyay, A; Sarkar, M. De; Bhowmick, AK. *J. Mater. Sci*, 2006, 41, 5981-5993.
[14] Sengupta, R; Bandyopadhyay, A; Sabharwal, S; Chaki TK; Bhowmick, AK. *Polymer* 2005, 46, 3343-3354.
[15] Gao, F; Chen S; Hull, JB. *J. Mater. Sci. Lett*, 2001, 20, 1807-1810.
[16] Usiki, A; Tukigase, A; Kato, M. *Polymer* 2002, 43, 2185-2189.
[17] Ray SS; Okamoto, M. *Polym. Sci*, 2003, 28, 1539-1641.
[18] Biswas, M; Ray, SS. *Adv. Polym. Sci*, 2001, 155, 167-221.
[19] Giannelis, EP; Krishnamoorti, R; Manias, E. *Adv. Polym. Sci*, 1999, 138, 107-147.
[20] Aranda, P; Ruiz-Hitzky, E. *Chem. Mater*, 1992, 4, 1395-1403.
[21] Wu, J; Lerner, MM. *Chem. Mater*, 1993, 5, 835-838.
[22] Strawhecker, KE; Manias, E. *Chem. Mater*, 2000, 12, 2943-2949.
[23] Chang, JH; Jang, TG; Ihn, KJ; Lee, WK; Sur, GS. *J. Appl. Polym. Sci*, 2003, 90, 3208-3214.
[24] Blumstein, A. *J. Polym. Sci. A*, 1965, 3, 2665-2673.
[25] Krishnamoorti, R; Vaia RA; Giannelis, EP. *Chem. Mater*, 1996, 8, 1728-1734.
[26] Vaia, RA; Ishii H; Giannelis, EP. *Chem. Mater*, 1993, 5, 1694-1696.
[27] Triantafillidis, CS; LeBaron, PC; Pinnavaia, TJ. *J. Solid State Chem.* 2002, 167, 354-362.
[28] 28. Su, S; Jiang, DD; Wilkie, CA. *Polym. Degrad. Stab*, 2004, 83, 321-331.
[29] Zhang, J; Jiang DD; Wilkie, CA. *Thermochimica Acta,* 2005, 430, 107-113.
[30] Xie, W; Xie, R; Pan, W.-P; Hunter, D; Koene, B; Tan, L.-S; Vaia, R. *Chem. Mater*, 2002, 14, 4837-4845.
[31] Patel, HA; Somani, RS; Bajaj, HC; Jasra, RV. *Appl. Clay Sci*, 2007, 35, 194-200.
[32] Wang, ZM; Chung, TC; Gilman JW; Manias, E. *J. Polym. Sci. Part B: Polym. Phys*, 41, 2003, 3173-3187.
[33] Chua YC; Lu, X. *Langmuir,* 2007, 23, 1701-1710.
[34] Costache, MC; Heidecker, MJ; Maias, E; Gupta RK; Wilkie, CA. *Polym. Degrad. Stab*, 2007, 92, 1753-1762.
[35] Chigwada, G; Wang, D; Wilkie, CA. *Polym. Degrad. Stab*, 2006, 91, 848-855.
[36] Wang D; Wilkie, CA. *Polym. Degrad. Stab*, 2003, 82, 309-315.
[37] Yao, H; Zhu, J; Wilkie CA; Morgan, AB. *Polym. Eng. Sci*, 2002, 42, 1808-1814.
[38] Zhang, J; Gupta RK; Wilkie, CA. *Polymer* 2006, 47, 4537-4543.
[39] Wang, K; Wang, L; Wu, J; Chen, L; He, C. *Langmuir* 2005, 21, 3613-3618.
[40] Persson, HHJ; Caseri, WR; Suter, UW. *Langmuir,* 2001, 17, 3643-3650.
[41] Ke, Y.C; Stroeve, P. *Polymer layered silicate and silica nanocomposites*; Elsevier Publisher, 2005; pp 52-56.
[42] Reddy, CR. Surface Modification and Characterization of Montmorillonite Clays-Catalytic Activity Studies on Selected Organic Reactions, Ph.D. Thesis, Visvesvaraya Technological University, India, July 2006.
[43] Vijayakumar, B. Synthesis of Industrially Important Esters using eco-friendly clay catalysts, Ph.D. Thesis, Visvesvaraya Technological University, India, September 2005.
[44] Lee, SY; Kim, SJ. *Clays Clay Miner*, 2002, 50, 435-445.

[45] Lagaly, G. *Clays Clay Miner*, 1982, 30, 215-222.
[46] Gregory, R; Smith DJH; Westlake, DJ. *Clay Miner*, 1983, 18, 431-435.
[47] Ballantine, JA; Davies, M; O'Neil, RM; Patel, I; Purnell, JH; Rayanakorn M; Williams, KJ. *J. Mol. Catal*, 1984, 26, 57-77.
[48] Choudhary, VR; Mantri, K; Jana, SK. *Catal. Commun*, 2001, 2, 57-61.
[49] Yadav, GD; Joshi, AV. *Clean Technol. Environ. Policy* 2002, 4, 157-164.
[50] Srinivas; Das, B. *J. Org. Chem*, 2003, 68, 1165-1167.
[51] Bharadwaj, RK; Mehrabi, AR; Hamilton, C; Trujillo, C; Murga, M; Fan, R; Chavira, A; Thompson, AK. *Polymer* 2002, 43, 3699-3705.
[52] Hwang, S.-H; Paeng, S.-W; Kim, J.-Y; Huh, W; Lee, S.-W. *Polym. Bull*, 2003, 49, 329-335.
[53] Wang, Y; Li, W. *React. Kinet. Catal. Lett*, 2004, 83, 195-203.
[54] Reddy, CR; Vijayakumar, B; Iyengar, P; Nagendrappa, G; Jai Prakash, B. S. *J. Mol. Catal. A: Chemical,* 2004, 223, 117-122.
[55] Vijayakumar, B; Iyengar, P; Nagendrappa, G; Jai Prakash, BS. *Ind. J. Chem. Technol*, 2004, 11, 565-568.
[56] Yadav, GD; Bhagat, RD. *Clean Technol. Environ. Policy,* 2005, 7, 245-251.
[57] Motokura, K; Fujita, N; Mori, K; Mizugaki, T; Ebitani, K; Kaneda, K. *J. Am. Chem. Soc*, 2005, 127, 9674-9675.
[58] Yadav GD; Lande SV. *Org. Process Res. Develop*, 2005, 9, 288-293.
[59] Reddy, CR; Iyengar, P; Nagendrappa, G; Jai Prakash, BS. *J. Mol. Catal. A: Chem*, 229, 2005, 31-37.
[60] Tsai, T.-Y; Li, C.-H; Chang, C.-H; Cheng, W.-H; Hwang, C.-L; Wu, R.-J. *Adv. Mater.* 2005, 17, 1769-1773.
[61] Yang, T; Wen, X.-D; Li, J; Yang, L. *Appl. Surf. Sci*, 2006, 252, 6154-6161.
[62] Williamson, DT; Schleinitz, HM; Hayes, R.A. *Polyester clay nanocomposites for barrier applications*, WO/2006/069131 (2006).
[63] Wallis, PJ; Gates, WP; Patti, AF; Scott, JL; Teoh, E. *Green Chem.* 2007, 9, 980-986.
[64] Benecke, HP; King, JL; Kawczak, AW; Zehander DW; Hirschl, EE. *Esters of 5-hydroxymethylfurfural and methods for their preparation,* WO2007/092569 A1(2007).
[65] Mittal, V. *J. Colloid Interface Sci*, 2007, 315, 135-141.
[66] Yadav GD; George, G. *J. Mol. Catal. A: Chem*, 2008, 292, 54-61.
[67] Zhang, Z; Ma, Y; Zhao, Y. *Syn. Lett*, 2008, 7, 1091-1095.

In: Polymer Nanocomposites: Advances in Filler Surface ISBN: 978-1-60876-125-8
Editor: Vikas Mittal 

*Chapter III*

# Grafting of Polymer Chains 'From' the Clay Surface

***Yongfang Yang and Hanying Zhao****

Department of Chemistry, Nankai University, Tianjin 300071, P. R. China

## Abstract

Several methods have been used in the preparation of exfoliated clay-polymer nanocomposites, e. g., in-situ polymerization of monomers, melt blending and solution blending in polar solvents. Of all the methods used in the preparation of nanocomposites, in situ polymerization method offers the best control over the polymer architecture and the structure of the composites. Various different polymerization methods were employed in the grafting of polymer chains from clay surface, including living radical polymerization method, living anionic polymerization method, ring-opening polymerization method and coordination polymerization method. The application of the polymerization methods in grafting of polymer chains on clay surface was summarized in this chapter. Besides grafting of polymer chains on clay surface, the grafting of polymer chains on the edge of clay layers was also reviewed.

**Keywords**: exfoliation, grafting, clay surface, polymerization methods, melt blending, solution blending.

## 3.1. Introduction

The preparation of polymer-layered silicate nanocomposites has generated intense interest in the materials science community over the past decade [1-3]. With relatively low loading of layered silicate, many properties of the composites improve significantly, such as mechanical properties, [4-6] thermal stability, [7-8] and flame retardance etc [9]. Clay is the

*Correspondence author: hyzhao@nankai.edu.cn

most commonly used layered silicate minerals. There are alkali or alkaline earth cations situated in the interlayer. Cationic surfactants such as alkylammonium or alkylphosphonium (onium) can exchange with the hydrated cations in the interlayer through ion-exchange reactions. Furthermore, there are silanol groups located on the edge of clay sheets. By reactions of silanol groups with coupling agents, a variety of functional groups such as chloro or alkoxy silanes could be attached to the edge of clay layers [10-13].

There are several methods to prepare polymer-clay nanocomposites, including in-situ polymerization, intercalation of polymers in solutions, direct intercalation of molten polymers and sol/gel technology. In in-situ polymerization the modified layered silicate is swollen by a liquid monomer or a monomer solution. The monomer molecules migrate into the galleries of the layered silicate so that the polymerization can occur between the intercalated sheets. After initiation by an organic initiator or catalyst fixed inside the interlayer before the swelling of the monomer, the long polymer chains bound to the clay galleries can be obtained. Of all the methods used in the preparation of polymer-silicate nanocomposites, in-situ polymerization offers the best control over both the polymer architecture and the final structure of the composite. It is also expected that nanocomposites made by in-situ block copolymerization will give significantly different morphologies and material properties compared with their counter-parts made by melt or solution intercalation. Polymer chains with one end attached to the clay layers are considered to be polymer brushes at the surface of clay layers. In situ polymerization from an initiator or catalyst attached to silicate layers is expected to be more akin to polymer brushes.

By ion exchange reaction different initiator or catalyst with quaternary ammonium terminal groups can be intercalated into the interlayer spacing of clay. Polymer chains on the clay surface can be obtained by free radical polymerization, anionic polymerization, ring-opening polymerization, atom radical polymerization, living/controlled polymerization such as atom transfer radical polymerization (ATRP), nitroxide-mediated polymerization (NMP), and reversible addition-fragmentation transfer (RAFT) polymerization.

## 3.2. Polymer Brushes Prepared by In-Situ Free Radical Polymerization

Because of easy operation and feasible to almost all vinyl monomers, free radical polymerization is widely used in the preparation of polymer-layered silicate nanocomposites. Fan and coworkers prepared a monocationic azobis(isbutyronitrile) (AIBN) derivative which can be attached to clay surfaces by electrostatic attraction. They prepared a polystyrene (PS) brush layer with a thickness of about 8 nm through solution free radical surface-initiated polymerization (SIP) from adsorbed clay nanoparticle on solid support substrates.

Both ellipsometry and surface plasmon spectroscopy (SPS) verified formation of distinct layers on the SIP protocol [14]. In order to prepare polymer-clay nanocomposite they synthesized two types of initiators: a symmetric bicationic azo compound, 2, 2’-azobis(isobutyramidine hydrochloride) (AIBA) and an asymmetric monocationic AIBN-analogue free radical initiator (Figure 1). They studied the structure of PS-clay nanocomposite prepared by free radical polymerization using the two different free-radical initiators modified clay. The authors found that the clay intercalated with monocationic initiator has a larger *d*-spacing and more ordered structure, and the SIP product of the bicationic initiator retained some intercalated structure while a completely exfoliated structure is achieved through the monocationic initiator. Size exclusion chromatography (SEC) showed that the SIP product of the monocationic initiator has a higher molecular weight (MW) for both free and bound polymer. The authors concluded that SIP on clay is more successful through the use of a monocationic initiator [15]. They also investigated the properties of clay surface bound initiators for the preparation of polymer-layered silicate nanocomposite materials: one for anionic polymerization and two for free radical polymerization. X-ray diffraction (XRD) results showed that the d-spacing increase of intercalated clay was dependent on the structure and functionality of different initiators. The difunctional initiator is less ideally packed than the monofunctional one [16].

Fan and coworkers prepared PMMA-clay nanocomposites by bulk, solution, and suspension SIPs using monocationic free radical initiator modified clay. XRD and transmission electron microscopy (TEM) showed that well-dispersed structure was accomplished by bulk and solution SIPs, while partially exfoliated structure in the suspension SIP product was observed. In general, the MWs of all surface bound polymers obtained by the three methods were lower compared to their corresponding free polymers. Using the same clay loading and initiator concentration, relatively higher MW products were achieved by suspension and bulk polymerizations compared to the solution method. The authors found that the polydispersities of the surface bound polymers were narrower, which is unusual for a free radical polymerization based on AIBN initiators [17].

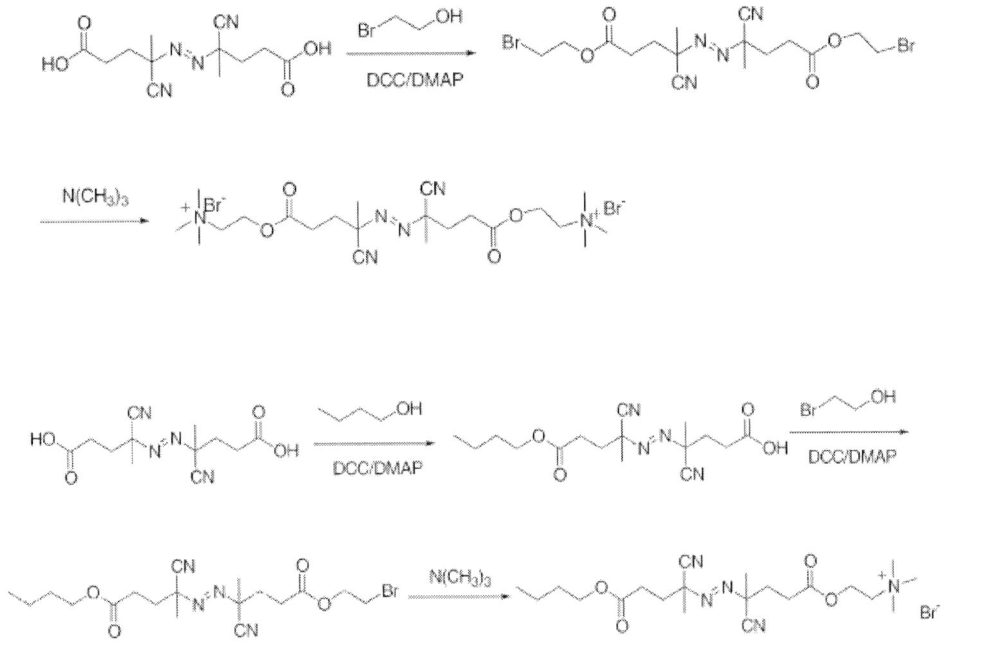


Figure 1. Synthetic scheme and structure of the bicationic free radical initiator and monocationic free radical initiator. Reproduced from Fan, Xia and Advincula by permission of American Chemical Society, USA.

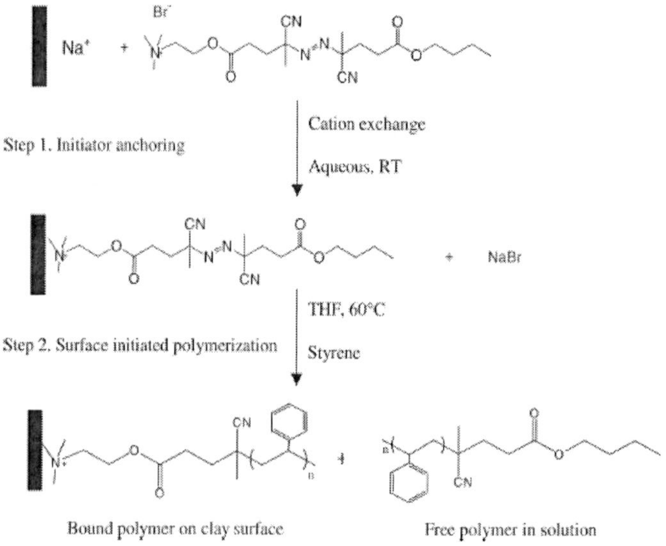


Figure 2. A scheme for the synthesis of PS-clay nanocomposites by free radical polymerization on the surfaces of clay layers. Reproduced from Vyazovkin, Dranca, Fan, and Advincula, American Chemical Society, USA.

Vyazovkin and coworkers prepared PS-clay nanocomposite by intercalating a monocationic free radical initiator into the interlayers of montmorillonite (MMT) clay and the subsequent solution SIP, where the chain growth was initiated in situ on clay surfaces (Figure 2) [18]. Thermogravimetry (TG) and differential scanning calorimetry (DSC) have been used to study the thermal and thermo-oxidative degradation of PS and PS-clay nanocomposite. The kinetic analysis demonstrates that the processes of degradation and relaxation have noticeably larger activation energies in the clay composite than in virgin PS. The larger energy barrier for the glass transition indicates lowering the molecular mobility and, therefore, increasing chemical and physical stability. For degradation, the increase in the activation energy also means an increase in stability due to retarding the process.

Uthirakumar and coworkers prepared the exfoliated PS/clay nanocomposites via in situ polymerization using a cationic radical initiator intercalated MMT hybrid [19]. The exfoliated structure resulted from the predominant intra-gallery polymerization over the extra-gallery polymerization owing to the anchored radical initiator inside the clay galleries. The thermal and mechanical properties of nanocomposites were improved drastically with low clay loading of up to 3wt%. However, the Young's modulus and tensile strength of PS/MMT nanocomposites were more significantly enhanced.

Zhang and coworkers reported preparation of PS and poly(methyl methacrylate) (PMMA) mixed polymer brushes on the surface of clay layers by using in situ free radical polymerization as shown in Figure 3[20]. Free radical initiator molecules with two quaternary ammonium groups at both ends were intercalated into the interlayer spacing of clay layers by cation exchange reaction. The amount of polymer brushes grafted on the surface of clay layers can be controlled by controlling the polymerization time. After in situ polymerization of styrene and methyl methacrylate, XRD result shows only a very weak and broad diffraction peak, which implies that the in situ polymerization created both intercalated structure and exfoliated structure in the nanocomposites. They also studied the nano-patterns of mixed copolymer brushes on the surface of clay layers. Their TEM image shows many PMMA nano-domains with an average diameter of 2 nm on clay surface, which indicates that because of immiscibility between PS and PMMA at a high temperature mixed polymer brushes tend to make phase separation forming nano-sized domains (Figure 4). The study on blends of PS and PMMA compatibilized by clay platelets with PS and PMMA mixed polymer brushes was also conducted. It was found that the intercalated clay particles tend to locate at the interface of two phases reducing the interfacial tension in the blends, as shown in (Figure 5). In the meanwhile, in the blends PMMA homopolymer chains tend to intercalate into clay layers. The driving force for the intercalation is the compatibility between homo-PMMA chains and PMMA brushes on the surface of clay layers.

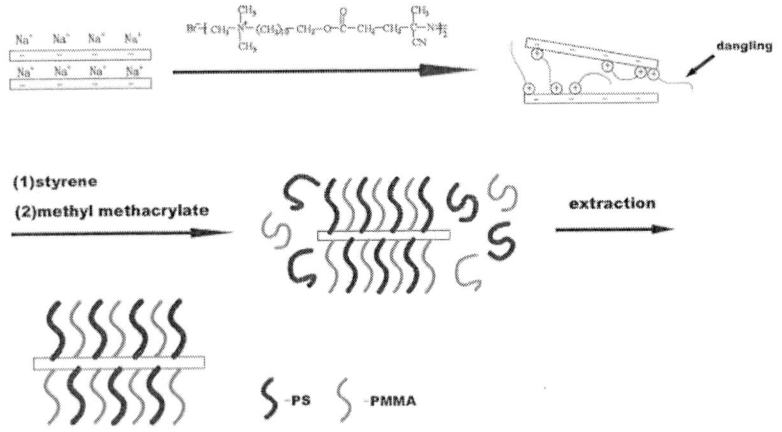


Figure 3. A schematic representation for the preparation of PS/PMMA mixed polymer brushes on the surface of clay layers. The specimens were stained by hydrazine hydrate and $OsO_4$, respectively. Reproduced from Zhang, Yang, Zhao and Zhao by permission of Wiley-VCH, Germany.

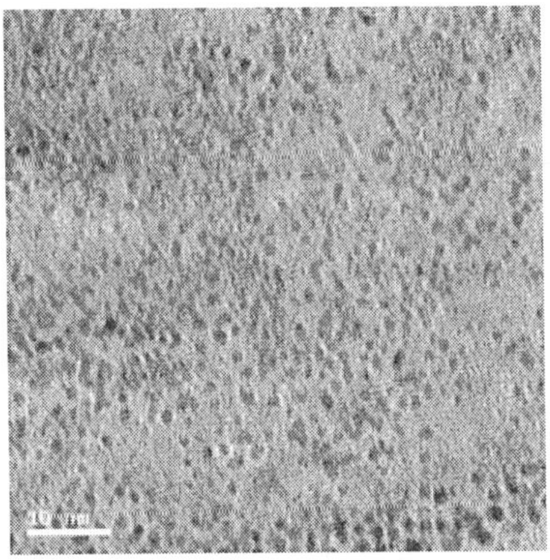

Figure 4. A high resolution TEM image of clay layer with PS and PMMA mixed brushes prepared from THF after annealing at 120 °C. To improve the contrast the PMMA phases were stained by exposing to hydrazine vapor for 10 h and $OsO_4$ vapor for 1.5 h. The specimens were stained by hydrazine hydrate and $OsO_4$, respectively. Reproduced from Zhang, Yang, Zhao and Zhao by permission of Wiley-VCH, Germany.

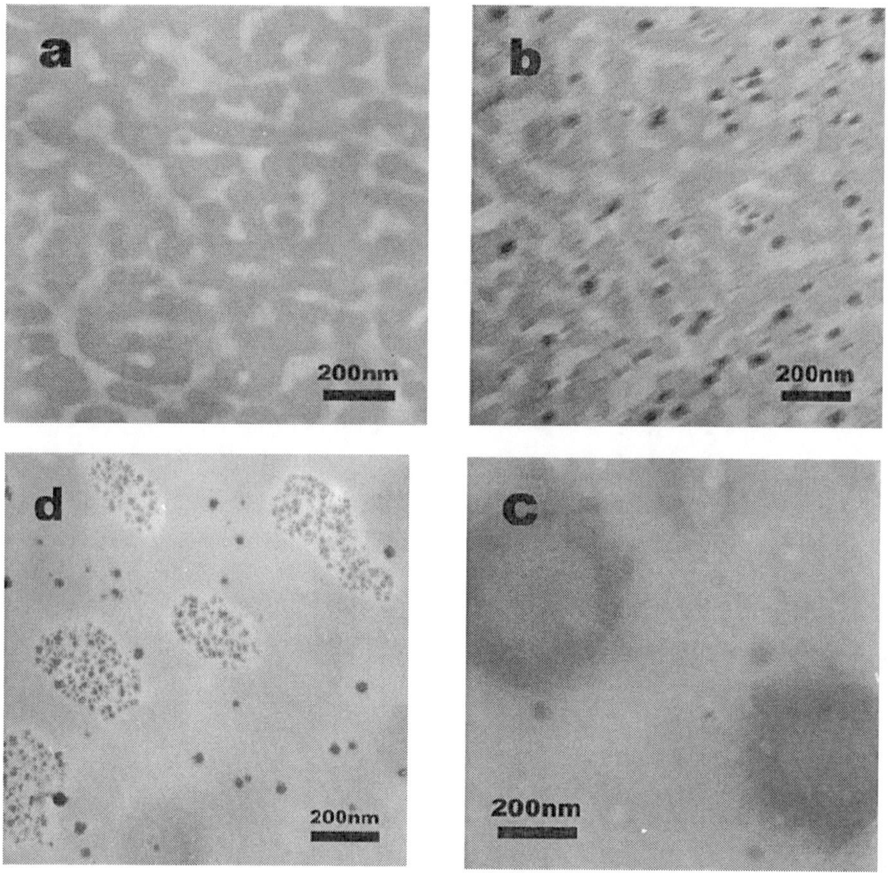


Figure 5. TEM images of PMMA/PS blends (33/67 w/w) with different amount of clay-PS6-PMMA10, (a) 0 wt %, (b) 1 wt %, (c) 3 wt %, and (d) 5 wt %. The specimenswere stained by hydrazine hydrate and $OsO_4$, respectively. Reproduced from Zhang,Yang, Zhao and Zhao by permission of Wiley-VCH, Germany.

In another study, PS and poly(2-(dimethylamino)ethyl methacrylate) (PDMAEMA) mixed polymer brushes on the surface of clay layers prepared by in situ free radical polymerization was reported by Yang and coworkers [21]. The clay layers with mixed homopolymer brushes can be used as stabilizer in suspension polymerization of styrene. PS colloid particles stabilized by clay layers with PS and PDMAEMA mixed polymer brushes were prepared by Pickering suspension polymerization. Figure 6 shows SEM images of PS colloid particles stabilized by varying amounts of PS-PDMAEMA/clay. With the increase of the amount of PS-PDMAEMA/clay added, the size of PS colloid particles decreases. Figure 7 shows a magnified TEM image of PS colloid particles. On the image clay layers on the surface of colloid particles can be observed clearly. The size of the irregular clay layers (indicated by an arrowhead) is roughly 130 - 150 nm.

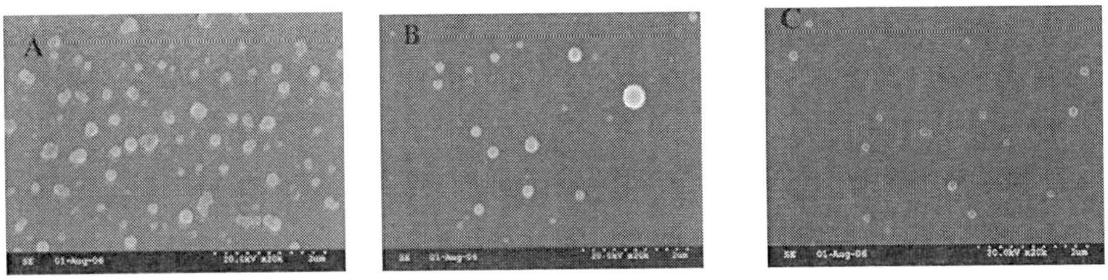


Figure 6. SEM images of PS colloid particles stabilized by varying amounts of PS-PDMAEMA/clay nanocomposites. The percentages of nanocomposite to styrene monomer are: (a) 0.5 wt %, (b) 1 wt %, (c) 1.5 wt %. Reproduced from Yang, Zhang, Liu, Li and Zhao by permission of Wiley-VCH, Germany.

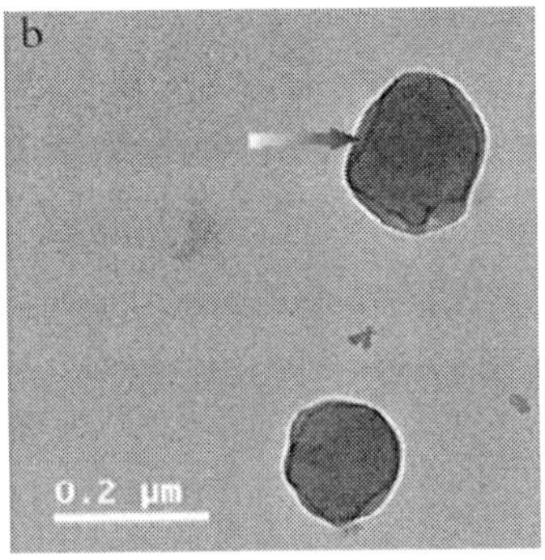


Figure 7. A TEM image of PS colloid particles stabilized by PS-PDMAEMA/clay Nanocomposites. Reproduced from Yang, Zhang, Liu, Li and Zhao by permission of Wiley-VCH, Germany.

Zhang and coworkers also reported the synthesis of PDMAEMA polyelectrolyte polymer brushes on the surfaces of clay layers by in situ free-radical polymerization, and application of the nanocomposites in the preparation of PMMA colloid particles by Pickering emulsion polymerization [22]. In the Pickering emulsion polymerization, PDMAEMA polymer brushes were used as a stabilizer and the remaining free-radical initiators on the surface of clay layers were used to initiate the polymerization of MMA. The structure of the clay-PDMAEMA composite and a schematic representation for the preparation of PMMA colloid particles by Pickering emulsion polymerization were shown in Figure 8. Because the initiation of the polymerization is on the surface of clay layers, the mechanism of the polymerization is different from that of a typical emulsion polymerization. The X-ray diffraction (XRD) results indicated that the intercalated structures of the clay layers were almost destroyed in Pickering emulsion polymerization, and clay layers with exfoliated structures were created. The surface of the colloid particles was analyzed by using X-ray photoelectron spectroscopy (XPS). The XPS results provide direct evidence that the clay layers with PDMAEMA chains cover the PMMA colloid particles. Because of a faceting effect caused by the rigidity of the clay platelets, PMMA colloid particles are not perfectly spherical (Figure 9.).

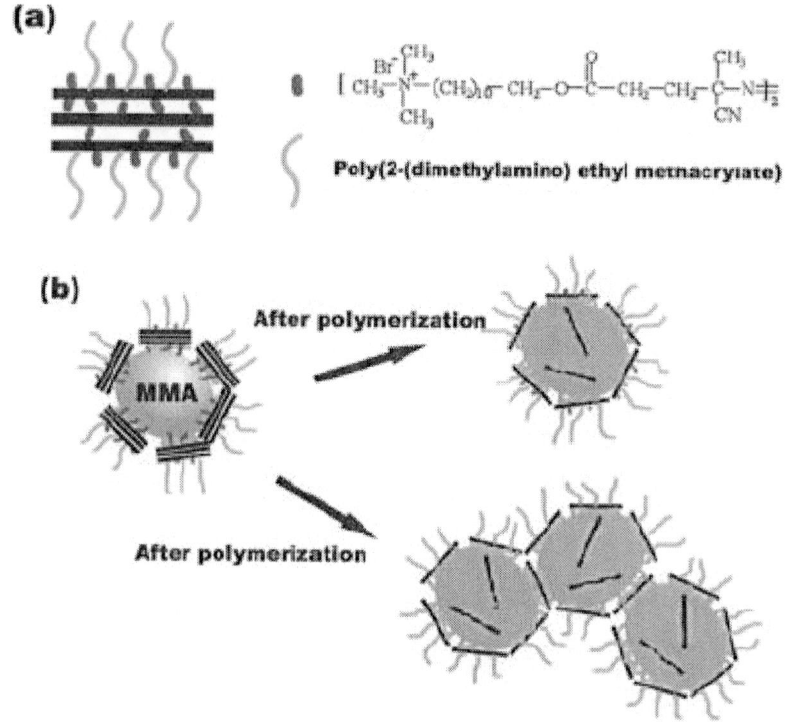


Figure 8. Schematic representations for the structure of clay-PDMAEMA composite (a) and preparation of PMMA colloid particles by Pickering emulsion polymerization (b). Reproduced from Zhang and Chen and Zhao by permission of Wiley-VCH, Germany.

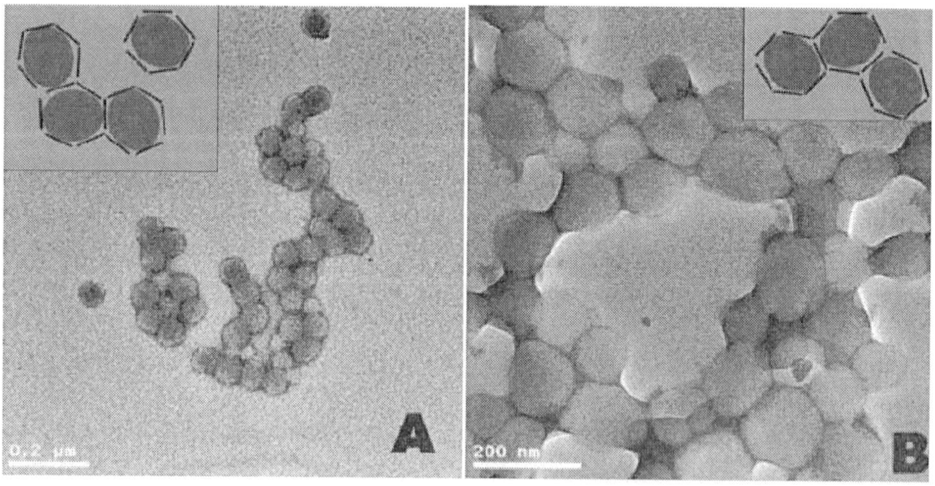


Figure 9. TEM images of PMMA colloid particles stabilized by (a) clay-PDMAEMA3 and (b) clay-PDMAEMA4. The molecular weights of PDMAEMA3 and PDMAEMA4 are 40K and 36K, respectively. Reproduced from Zhang and Chen and Zhao by permission of Wiley-VCH, Germany.

Not only free-radical initiators but also monomer molecules can be ion-exchanged onto surface of clay layers, and polymer-clay nanocomposites can be prepared by in situ free radical polymerization. For example, in a research vinyl monomer-montmorillonite intercalate (VM-MMT) was prepared by exchanging the mineral cations of MMT by vinylbenzyltrimethylammonium chloride. The resulting VM-MMT material rendered the mineral organophilic and having polymerizable moieties directly bonded to the lamellar surface of the mineral. Radical polymerization of styrene in the interlayers of MMT was conducted, and which resulted in PS-MMT intercalate materials [23, 24].

In another research Fu and coworkers synthesized a reactive cationic surfactant vinylbenzyldimethyldodecylammonium chloride-VDAC [25]. VDAC-intercalated MMT particles were easily dispersed and swollen in styrene monomer. PS–clay nanocomposites were prepared by free radical polymerization of styrene with dispersed clay. XRD and TEM indicated that exfoliation of MMT was achieved. PS–clay nanocomposites have higher dynamic modulus than pure PS.

Guymon and coworker reported a new polymerizable organoclays [26]. They synthesized methacrylated (C14MA and PM1) and thiol (PSH2) quaternary ammonium surfactants and modified clay surfaces through typical cation exchange procedures (Figure 10). They found that the polymerizable organoclays exfoliate more readily in a number of monomer systems as compared to non-polymerizable organoclays.

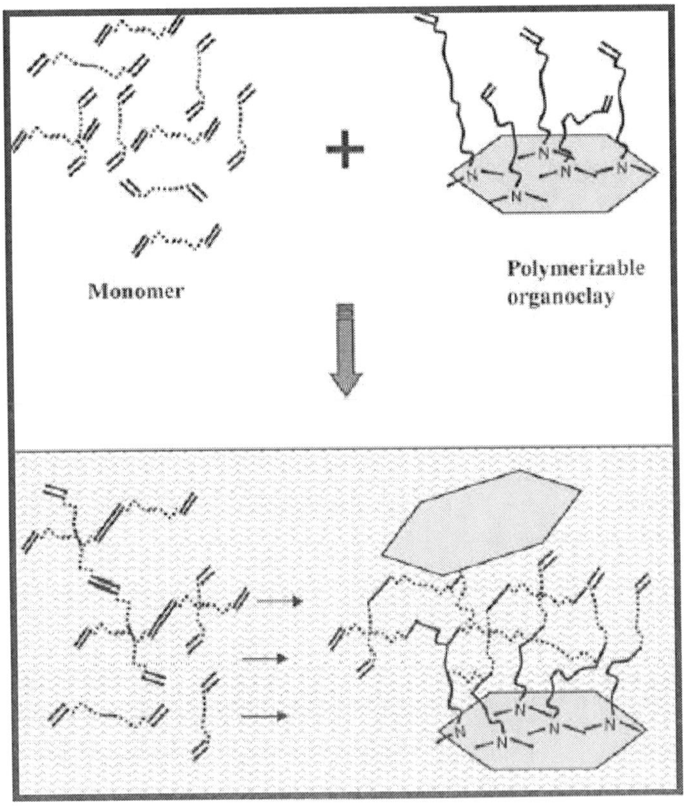


Figure 10. Schematic of Monomer Diffusion into Organoclay Interlayer. Reproduced from Owusu-Adom and Guymon, American Chemical Society, USA.

The position of the reactive functional group and the type of functionality influence the degree of exfoliation. Methacrylate functional groups located close to the clay surface result in intercalated organoclay domains with small aggregate sizes. When the reactive moiety is located away from the clay surface, creating a more accessible functional group, organoclay exfoliation occurs even before photo-polymerization in a variety of monomers. For thiolated organoclays, enhanced exfoliation occurs through copolymerization of thiol functional groups and acrylate monomer during photo-polymerization. Incorporating reactive organoclays lead to higher moduli and glass transition temperatures in the nanocomposites while also facilitating higher photo-polymerization rates compared to non-polymerizable analogues.

## 3.3. Polymer Brushes Prepared by In-Situ Controlled/Living Radical Polymerization (C/LRP)

In the past decade, C/LRP techniques have been used widely in the preparation polymers with different architectures, predetermined molecular weights, and low molecular weight distributions. C/LRP techniques are also used in the preparation of polymer brushes on layered silicate. Polymer chains prepared on the clay surface by ATRP was first reported by Henrik Böttcher [27]. The ATRP-initiating head group is was introduced through esterification of 11-bromoundecan-1-ol with 2-Bromoisobutyryl bromide quite easily. By reaction of 11'-Bromoundecanyl-2-bromoisobutyrate with trimethylamine, 11'-(N, N, N-trimethylammonium bromide)undecanyl-2- bromoisobutyrate was obtained, and which can could be ion-exchanged onto Na-montmorillonite (Figure 11a). Polymerization of methyl methacrylate was carried out from ATRP initiator-modified layered silicate. XRD measurements at different times of conversion demonstrate an increase of delamination as a result of increasing chain length of the grafts, and which finally results in a fully dispersed polymer silicate nanocomposite.

Zhao and coworkers prepared polystyrene, polymethylmethacrylate, poly(butyl acrylate), and poly(styrene-*block*-butylacrylate) block copolymer-silicate nanocomposites by in situ ATRP [28-32]. They also studied the formation of nano-patterns by block copolymer brushes on the surface of exfoliated or intercalated clay layers after treatment in different solvents. For poly(styrene-*block*-butyl acrylate block copolymer brushes, a uniform collapsed brush layer was observed after treatment in THF. However, after treatment in acetone, a selective solvent for PBA, wormlike surface aggregates were observed.

After treatment in methanol, a precipitant for both of the blocks, surface micelles as well as wormlike aggregates could be observed. Furthermore, the polymer brushes tend to aggregate together and change their nano-patterns at an elevated temperature. Yang and coworkers synthesized amphiphilic poly(methyl methacrylate- *block*- 2-(dimethylamino)ethyl methacrylate) (PMMA-*b*-PDMAEMA) block copolymer brushes on the surface of clay layers by in situ ATRP [33]. They also studied the application of the nanocomposite in Pickering emulsion. After being dispersed in an O/W system, the clay layers with block copolymer brushes self-assembled on the oil surface (Figure 12). Because MMA monomer is a good solvent for both of PMMA and PDMAEMA blocks, the polymer brushes stretch away from the clay surface into the oil phase, and in water phase the block copolymer brushes form surface micelles. The different conformations of block copolymer brushes at the interface reduce the surface energy between the two fluids and stabilize the oil droplets in aqueous phase. Clay layers with block copolymer brushes can be used as stabilizer in suspension polymerization of MMA, and PMMA colloid particles armored by clay nanocomposites were prepared.

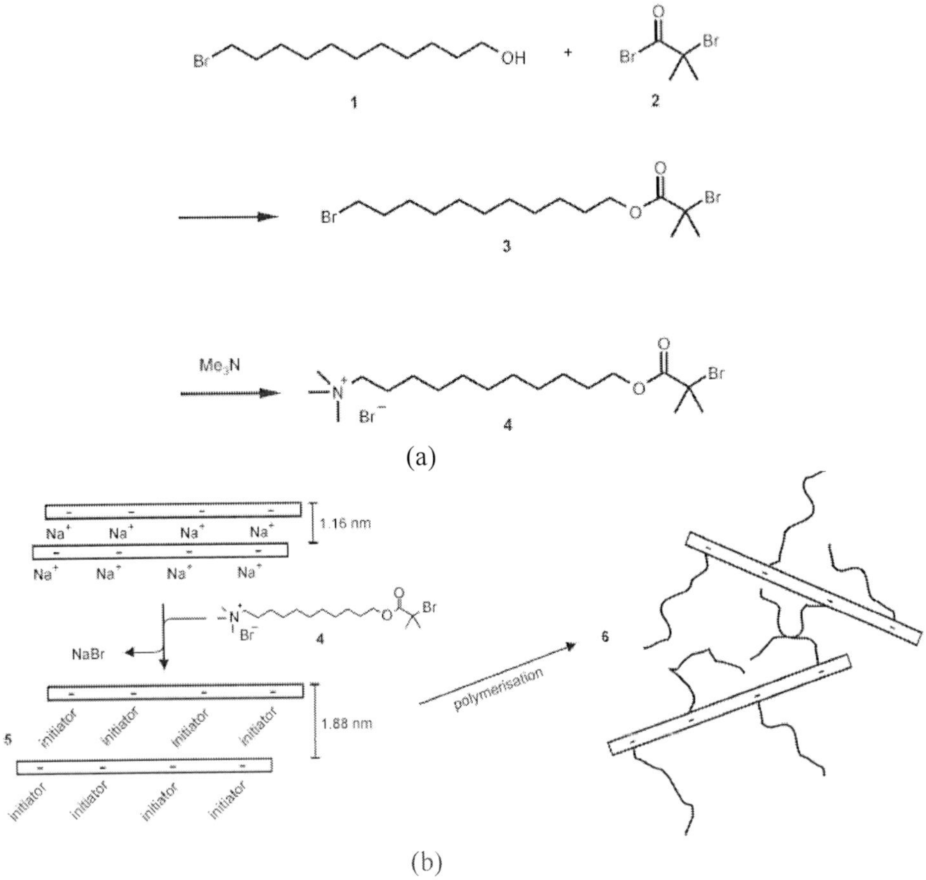


Figure 11. Schemes for the synthesis of ATRP initiator (a), and ATRP graft polymerization of methyl methacrylate (MMA) from initiator-modified montmorillonite (b). Reproduced from Böttcher, Hallensleben, Nuß, Wurm, Bauer and Behrens by permission of The Royal Society of Chemistry, UK.

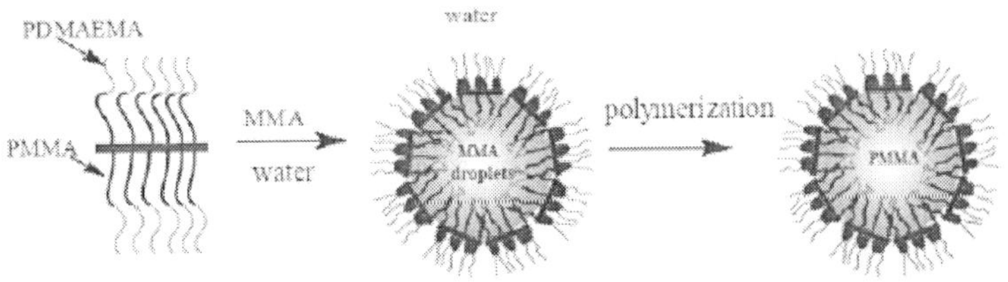


Figure 12. A schematic representation of the stabilization of MMA droplets and PMMA colloid particles by clay layers. Reproduced from Yang, Liu, Zhang, Li and Zhao, American Chemical Society, USA.

RAFT polymerization was also used in the preparation of polymer brushes on the surface of clay layers. Di and coworkers reported preparation of exfoliated polymer/silicate nanocomposites with silicate contents from 7 wt% to 63 wt % by in situ RAFT polymerization using a silicate-anchored dithiocarbarmate photoiniferter (Figure 13) [34]. Nanocomposites containing polystyrene, poly(methyl methacrylate), poly(*tert*-butyl methacrylate), and poly(methyl methacrylate-*b*-styrene) of controlled molecular weight, architecture, and polydispersity were prepared. XRD results showed that the silicate layers were randomly distributed in the polymer matrix. Not only exfoliated silicate layers with random orientations but also regions containing a stacking of 4-6 parallel silicate layers with interlayer spacing of 6-8 nm could be observed.

Weimer and coworkers reported anchoring of nitroxide-mediated free radical polymerization (NMP) initiator inside the galleries of layered silicate hosts followed by intercalation and polymerization of styrene (Figure 14) [35]. After polymerization dispersed PS-silicate nanocomposite was obtained. They demonstrated that intergallery polymerization via a silicate-anchored initiator is a viable approach to direct synthesis of dispersed silicate nanocomposites. In addition, critical polymer characteristics such as MW and polydispersity are controlled, and block copolymer formation appears possible.

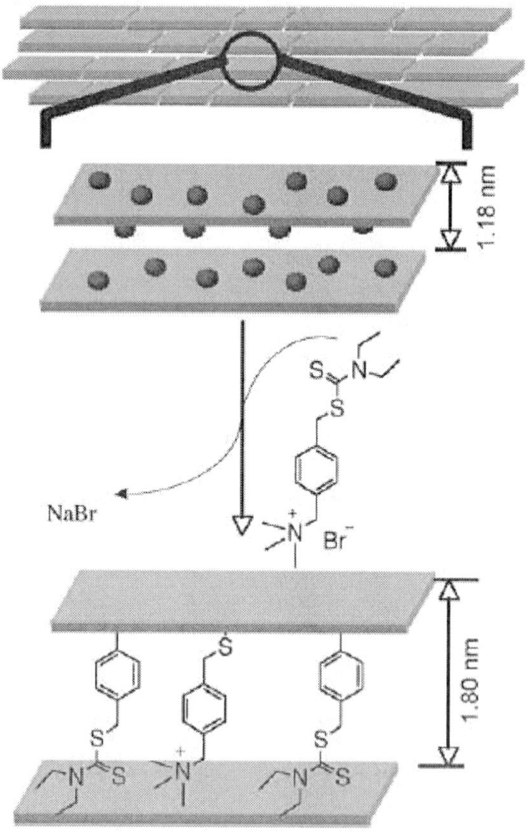


Figure 13. Representation of the RAFT-related photoiniferter intercalation into the silicate layers via ion exchange. Reproduced from Di and Sogah, American Chemical Society, USA.

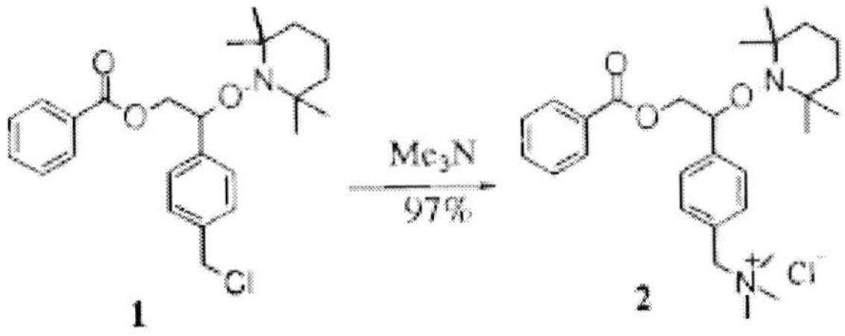


(a)

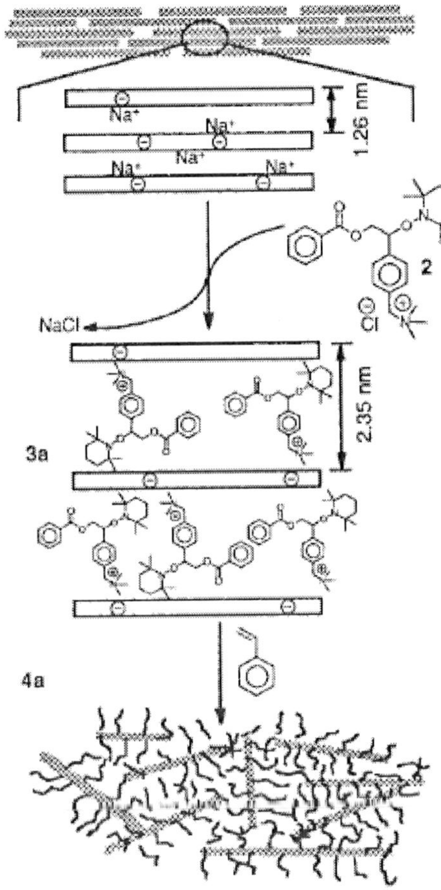


(b)

Figure 14. Schemes for (a) the synthesis of initiator and (b) preparation of PS-silicate nanocomposite by nitroxide-mediated free radical polymerization. Reproduced from Weimer, Chen, Giannelis and Sogah, American Chemical Society, USA.

Di and coworkers designed a multifunctional initiator that contained a benzyltrimethylammonium group as an anchoring site and initiating sites for two different polymerization mechanisms, namely, a primary alcohol functionality for living ring opening polymerization of heterocyclic ethers and lactones, and an alkoxyamine for nitroxide-mediated living free radical polymerization (Figure 15) [36]. Poly(styrene-*b*-caprolactone)/silicate nanocomposites with exfoliated morphology were prepared via one-pot, one-step in-situ block copolymerization involving simultaneous living free radical polymerization of styrene and anionic ring-opening polymerization of caprolactone in the presence of a bifunctional initiator-anchored MMT. Anchoring the block copolymer at the junction between the two blocks would make it possible for each block to more readily diffuse into and interact or entangle with a chemically similar domain in a phase separated blend for which the block copolymer is used as a compatibilizing agent. The author think it to be more effective in compatibilizing immiscible polymer blends and reinforcing their interfaces than would be expected from block copolymer/silicate nanocomposites in which one end of the diblock copolymer is restricted, leaving only the free end of the other block. XRD and DSC results showed that the PCL segment existed in a crystalline state in the nanocomposite.

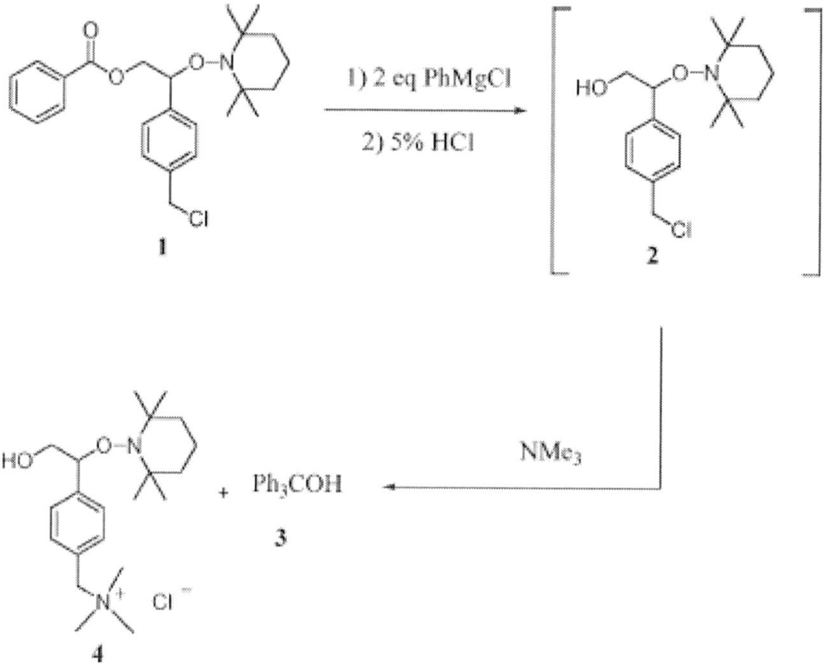


(a)

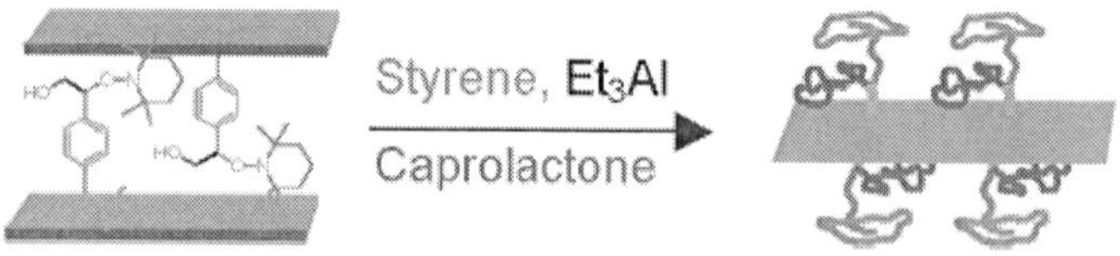


(b)

Figure 15. (a) Synthesis of the Bifunctional Initiator, (b) preparation of diblock copolymer/silicate nanocomposites based on ring-opening polymerization and nitroxide-mediated free radical polymerization. Reproduced from Di and Sogah, American Chemical Society, USA.

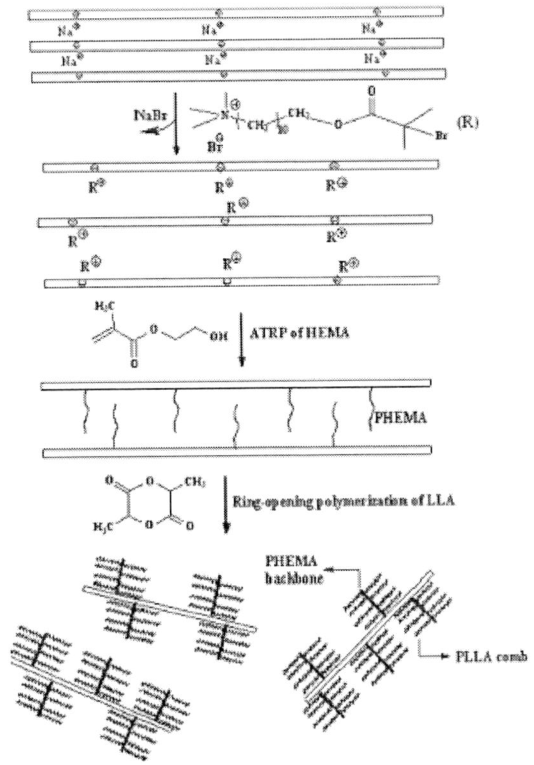


Figure 16. A schematic representation for the preparation of PLLA comb polymer brushes on the surface of clay layers. Reproduced fromYang, Wu, Li, Liu, Cheng and Zhao by permission of Elsevier Science Ltd., UK.

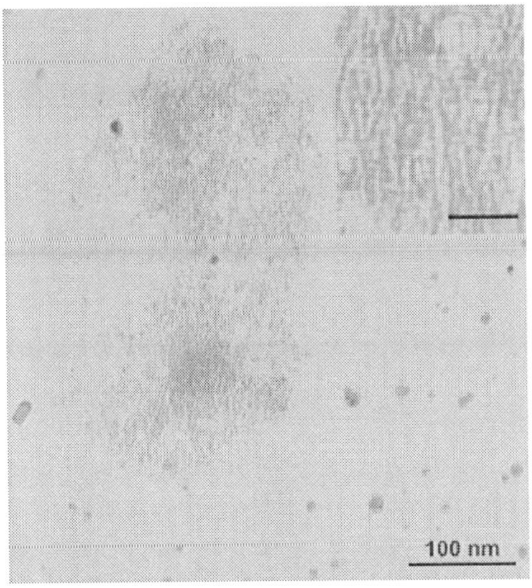


Figure 17. A TEM image of PLLA-PHEMA/clay nanocomposite prepared from a dispersion of the nanocomposites in $CHCl_3$ solution. In order to increase the contrast, the specimen was stained by hydrazine hydrate and OsO4, respectively. The inset is a magnification of a part of isolated clay layer with block copolymer brushes and the scale bar represents 50 nm. Reproduced fromYang, Wu, Li, Liu, Cheng and Zhao by permission of Elsevier Science Ltd., UK.

Yang and coworkers prepared poly(L-lactide) (PLLA) comb polymers on poly(hydroxyethyl methacrylate) (PHEMA) backbone on the surface of clay layers by a combination of in situ atom transfer radical polymerization (ATRP) and ring-opening polymerization [37]. An ATRP initiator with a quaternary ammonium salt end group was intercalated into the interlayer spacing of clay. PHEMA polymer brushes on the surface of clay layers were prepared by in situ ATRP. PLLA comb polymers on PHEMA backbone were prepared by ring-opening polymerization (Figure 16). In Figure 17, a TEM image showed clearly the structure of the comb polymer brushes on the surface of an exfoliated clay layer. The inset in Figure 17 is a magnification of a part of isolated clay layer. The tethered comb polymer brushes tend to aggregate together on the clay surface.

## 3.4 Polymer Brushes Prepared by In-Situ Anionic Polymerization

It is very difficult to conduct anionic polymerization in a system with layered silicate, because the hydrophilic lamella hosts inevitably contain moisture, which can terminate a propagating living anion. However, by carefully controlling the polymerization conditions and after treatment of layered silicate, anionic polymerization is feasible in the preparation of the polymer brushes. Zhou and coworkers used a 1,1-diphenylethylene (DPE) derivative as an initiator precursor in anionic polymerization (Figure 18a)[38, 39]. DPE was modified to be an organic cation by functionalizing with a triethylammonium bromide group, and then the modified DPE was immobilized to clay surface by ion-exchange. At a high temperature and under a high vacuum, styrene was successful ionic- polymerized on clay layers (Figure 18b). The molecular weight and the molecular weight distribution data of the detached polymer indicate a *living anionic* mechanism.

They found that the bound polymer has a much lower molecular weight and broader distribution than the polymer formed in solution. They believed that the tight intergallery spacing between clay platelets may had prohibited the styrene monomer from reaching the DPE anionic sites in the interlayers when the monomer was added to the suspension of clay. They also found that the initiation/chain growth efficiency inside clay interlayer spaces was lower than that on clay perimeter surfaces.

## 3.5 Polymer Brushes Prepared by In-Situ Ring Opening Polymerization

Ring opening polymerization is widely used in the preparation of biocompatible polymers. In-situ ring-opening polymerization on the surface of layered silicate provides an efficient way to the synthesis of biocompatible polymers-silicate nanocomposites. A Toyota research group first reported the ability of α,ω-amino acid (COOH–(CH2)$_{n-1}$–NH2$^{+}$, with $n$=2, 3, 4, 5, 6, 8, 11, 12, 18) modified Na$^{+}$-MMT to be swollen by ε-caprolactam monomer at 100°C and subsequently initiate ring opening polymerization to obtain PA6/MMT nanocomposites [40]. Figure 19 represents the conceptual view of the swelling behavior of α,ω-amino acid modified Na+-MMT by ε-caprolactam [41]. XRD and TEM showed exfoliated or intercalated structures were obtained depending on the amount of organic modified clay in the system.

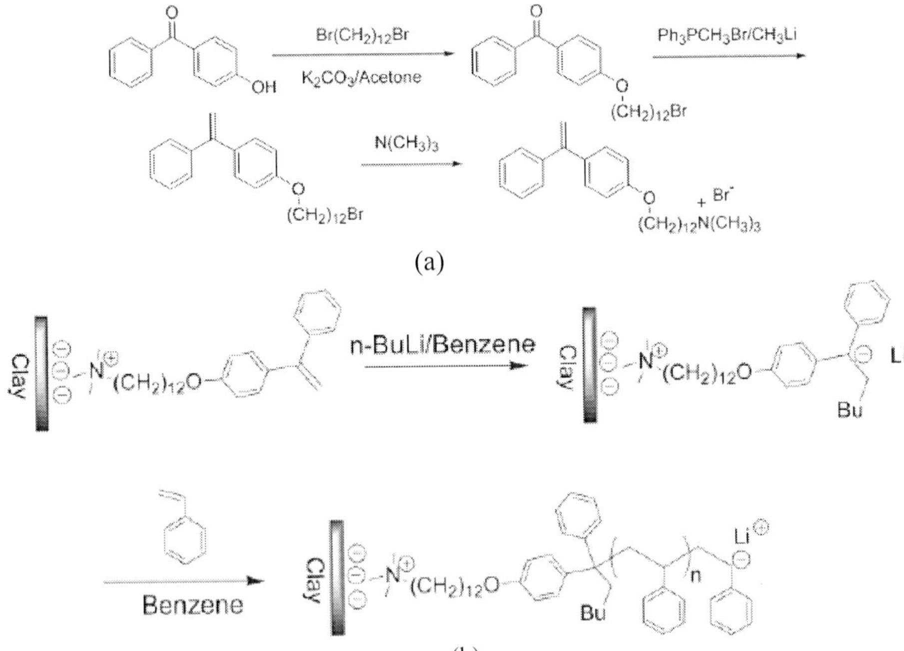


Figure 18. (a) Synthetic scheme of the cationic (charged) 1,1-diphenylethylene (DPE) derivative initiator, and (b) Schematic illustration of living anionic surface-initiated polymerization (LASIP) from clay surfaces. Reproduced from Fan, Zhou, Xia, Cristofoli, Mays and Advincula by permission of American Chemical Society, USA.

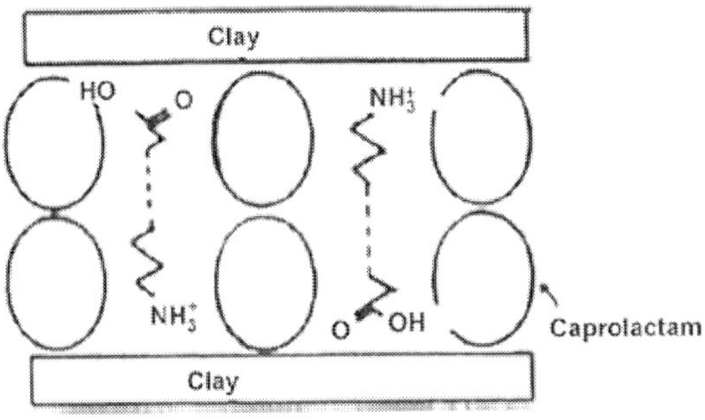


Figure 19. Swelling behavior of α,ω-amino acid modified MMT by ε-caprolactam. Reproduced from Usuki, Kawasumi, Kojima, Okada, Kurauchi and Kamigaito by permission of Materials Research Society, USA.

Using both pristine MMT and aminododecanoic acid modified MMT, PCL nanocomposites were prepared by Pantoustier and coworkers [42,43]. The polymerization of CL with pristine and modified MMT gives PCL with a molar mass of 4800 and 7800 g/mol, respectively, and narrow distributions. They found no CL polymerization under the same experimental condition without MMT. These results demonstrate the ability of MMT to catalyze and control CL polymerization, at least in terms of molecular weight distribution which remains remarkably narrow. The authors assumed that CL is activated through interaction with the acidic site on the clay surface.

The same researchers also prepared PCL nanocomposites using a protonated aminolauric acid modified Na+-MMT by ring opening polymerization at 170°C for 48 h. XRD patterns of the modified clay after contact with ε-caprolactone at room temperature do not show any significant increase in the layer spacing. The authors assumed that the monomer intercalates in the gaps between the aminolauric acid chains so that no gallery expansion could be seen. Another possibility may be that intercalation of the monomer occurs only during the heating step of the solution. After polymerization, XRD patterns of the obtained composites did not show a diffraction peak, indicating that exfoliation was achieved [44].

### 3.6. Polymer Brushes Initiated by Intercalated Catalysts

Mariott and coworkers reported the first synthesis and characterization of stereochemically controlled PMMA-silicate nanocomposites by using the intergallery-anchored metallocenium cations [45]. The intergallery-anchored metallocenium cations was prepared via a non-cation-exchange approach involving protonolysis of the metallocene dimethyl with Me(HT)2NH+/MMT, a montmorillonite (MMT) clay modified by methyl bis(hydrogenated tallow alkyl) ammonium. These intercalated metallocene catalysts allow for the in situ synthesis of nanocomposites comprising delaminated silicate nanoplatelets dispersed in the atactic, isotactic, or syndiotactic PMMA matrixes. The access to such nanocomposites provides possibility for fundamental studies of polymer stereochemistry effect on nanocomposite properties.

Jin and coworkers used organic salts with hydroxyl groups to modify MMT [46]. The hydroxyl groups offer facile reactive sites for anchoring catalysts and silicate layers. A Ti based Ziegler-Natta catalyst was fixed at the inner surface of MMT and ethylene was polymerized. Figure 20 represents the TiCl4 fixation mechanism between silicate layers of MMT–OH. WAXD analysis revealed exfoliation of MMT layers in both the powdery reaction product and the HDPE matrix.

In a research by Shin and coworkers montmorillonite was intercalated by triisobutylaluminum and ω-undecylenyalcohol [47]. The intercalation process allows the transition metal catalyst and the activator methylaluminoxane to enter the clay galleries and conduct the polymerization of ethylene. The formation of polyethylene inside the galleries leads to the exfoliation of layered silicates. In the research two types of polymerization are possible: (a) the homopolymerization of ethylene, and (b) the copolymerization of ethylene and the vinyl ends of alcohol modifier connected to the surface during intercalation.

Tudor and coworkers prepared PP/clay nanocomposites with a synthetic hectorite modified by the transition metal catalyst [48]. The synthetic hectorite was treated with methylaluminoxane (MAO) to remove the acidic protons and the transition metal catalyst was anchored to the interlayer. Using a synthetic fluorinated mica-type layered silicate deprived of protons in the galleries, the catalyst was intercalated directly within the silicate layers without the need of MAO treatment. Using this method Sun and Garces prepared PP/clay nanocomposites by in-situ polymerization with metallocene/clay catalysts [49].

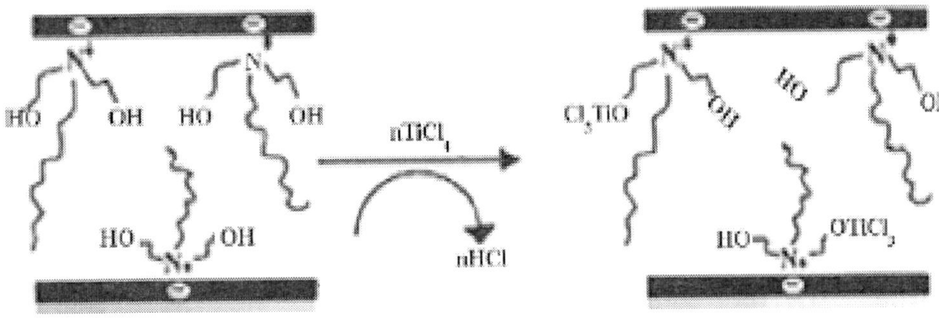


Figure 20. Mechanistic representation of the fixing of TiCl4 between the silicate layers of MMT–OH. Reproduced from Jin, Park, Im, Kwak and Kwak by permission of Wiley-VCH, Germany.

## 3.7 Polymer Chains on the Edges of Clay Sheets Prepared by In-Situ Polymerization

Considerable efforts to date have been devoted to exchange inorganic cations on the surface of clay with organic aliphatic quaternary ammonium to prepare polymer-clay nanocomposites. Besides ion exchanging, another modification is reaction of chloro or alkoxy silanes with the silanol groups located on the edges of the clay sheets. By reaction of silano groups with coupling agents, a variety of functional groups or initiators could be attached to the edges. Wheeler and coworkers reported the synthesis of polymer chains grown from covalently bound ATRP initiators from Laponite clay [50]. They prepared edge modified Laponite clay using alkoxy silanes possessing additional reactive groups such as primary amines, methacrylates, benzophenones, and tertiary bromines [51]. The silanol groups were reacted with aminopropyltrimethoxy silane (APS), aminopropyldimethylethoxysilane (APES), or trimethoxypropylsilane (TPS). Attaching primary amines to the edges of clay opens up a wide range of possible functionalization reactions. Thus, APS-treated clay was reacted to give attached methacrylate, benzophenone, and tertiary bromine groups capable of polymerization, photoinitiation, and ATRP initiation, respectively. With such structures, polymer chains can be grafted to or grown from clay edges, creating star-like or fringed polymers with a nano-sized inorganic core [52].

Using edge modified Laponite clay, Wheeler and coworkers explored two polymer attachment methods, one through reaction of a methacrylate compound and the clay's silanol group followed by in situ free-radical polymerization of methyl methacrylate (MMA), and the other through attachment of an ATRP initiator followed by brush polymerization (Figure 21). The resulting nanocomposites were optically transparent and homogeneous. The two methods yield different structures. Attaching methacrylate groups to the clay could lead to crosslinked structures surrounding the clay and could also link the sheets together, whereas ATRP polymerization should produce linear chains attached to an inorganic core [53].

Herrera and coworkers report the chemical modification of Laponite clay minerals using monofunctional ç-methacryloxy propyl dimethyl ethoxysilane (ç-MPDES) and trifunctional ç-methacryloxy propyl trimethoxysilane (ç-MPTMS) molecules, respectively. The functionalized clay particles were dispersed into water and engaged further in a free radical emulsion polymerization process. By using silane coupling agents, they produced colloidal nanocomposites with covalent grafting between the exfoliated clay sheets and the polymeric matrix. Moreover, polymer attachment on the inorganic surface since the beginning of the emulsion polymerization process ensures polymer growth from the seed surface [54].

Wu and coworkers prepared hydrophobic PS brushes on the edges of clay sheets by using edge modification and in situ ATRP [55]. Owing to hydrophilic clay surface and hydrophobic PS brushes, the clay-PS particles can be used as stabilizer in Pickering suspension polymerization of styrene. Due to the negatively charged clay particles on the surface, the zeta potential of the PS colloidal particles was negative. Positively charged poly(2-vinyl pyridine) (P2VP) chains were adsorbed to the surface of PS colloidal particles in aqueous solution at a low pH value, and gold nanoparticles were prepared in P2VP brushes (Figure 22).

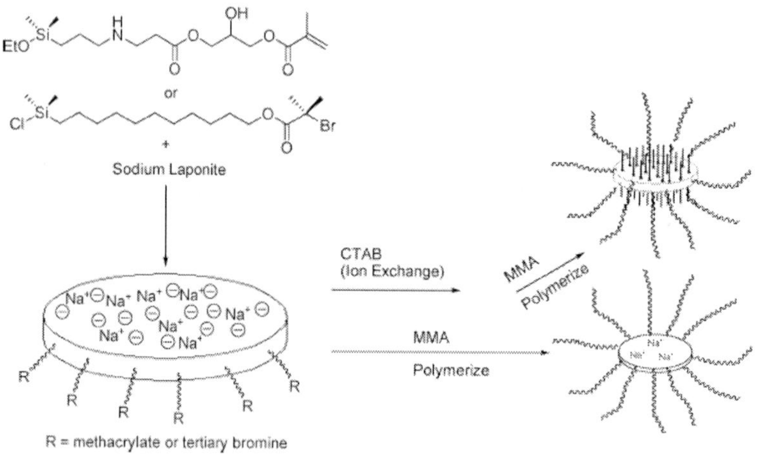


Figure 21. Scheme showing process of covalently attaching polymer to Laponite. Reproduced from Wheeler, Wang and Mathias by permission of American Chemical Society, USA.

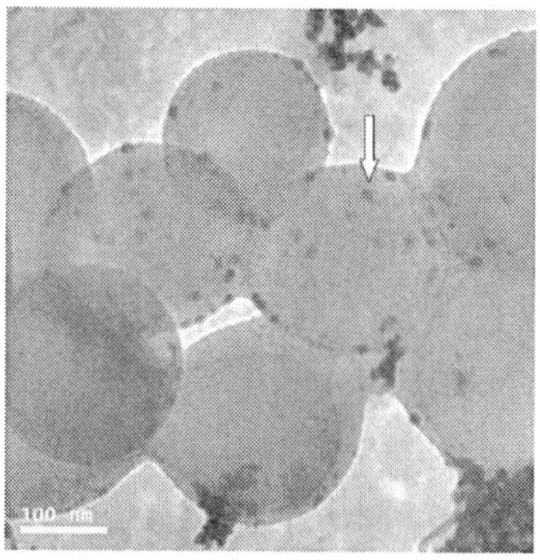

Figure 22. A TEM image of PS colloidal particles with gold nanoparticles on the surface. The PS colloidal particles stabilized by clay-PS were coated with poly(2-vinyl pyridine) (P2VP) in aqueous solution at pH 3.0 and gold nanoparticles were prepared in P2VP brushes. The specimens were stained by hydrazine hydrate and OsO4, respectively. Reproduced from Wu, Zhang and Zhao by permission of Wiley-VCH, Germany.

## 3.8. Summary

Polymer chains from clay surface can be prepared by in-situ polymerization when initiator, catalyst or monomer molecules are attached to the interlayer or the edges of the clay sheets by ion exchange or silane coupling reaction. In situ polymerization provides an efficient way to prepare polymer/clay nanocomposites with exfoliated or a mixture of exfoliated and intercalated structures. Polymer chains on the surface of clay layers can be used to modify the hydrophilic property of the nature clay and to improve the mechanical properties of the nanocomposites.

## References

[1] Usuki, A; Kawasumi, M; Kojima, Y; Okada, A; Kurauchi, T; Kamigaito, O. *J. Mater. Res*, 1993, 8, 1174.

[2] Usuki, A; Kojima, Y; Kawasumi, M; Okada, A; Fukushima, Y; Kurauchi, T; Kamigaito, O. *J. Mater. Res*, 1993, 8, 1179.

[3] Kojima, Y; Usuki, A; Kawasumi, M; Okada, A; Fukushima, Y; Kurauchi, T; Kamigaito, O. *J. Mater. Res*, 1993, 8, 1185.

[4] Lan, T; Pinnavaia, T. *J. Chem. Mater*, 1994, 6, 2216.

[5] Wang, Z; Pinnavaia, T. *J. Chem. Mater*, 1998, 10, 1820.

[6] Okada, A; Usuki, A. *Mater. Sci. Eng*, 1995, C3, 109.

[7] Messersmith, P; Giannelis, E. P. *Chem. Mater*, 1994, 6, 1719.

[8] Kojima, Y; Usuki, A; Kawasumi, M; Okada, A; Kurauchi, T; Kamigaito, O. *J. Polym. Sci. Part. A: Polym. Chem*, 1993, 31, 1755.

[9] Gilman, JW; Jackson, CL; Morgan, AB; Harris, R; Manias, E; Giannelis, EP; Wuthenow, M; Hilton, D; Phillips, SH. *Chem. Mater*, 2000, 12, 1866.

[10] Swartzen-Allen, SL; Matijevic, E. *Chem. Rev*, 1974, 74, 385.

[11] Wheeler, PA; Wang, J; Baker, J; Mathias, L. *J. Chem. Mater*, 2005, 17, 3012.

[12] Wheeler, PA; Wang, J; Mathias, L. *J. Chem. Mater*, 2006, 18, 3937.

[13] Herrera, NN; Letoffe, J; Putaux, J; David, L; Bourgeat-Lami, E. *Langmuir* 2004, 20, 1564.

[14] Fan, X; Xia, C; Fulghum, T; Park, MK; Locklin, J; Advincula, R.C. *Langmuir* 2003, 19, 916.

[15] Fan, X; Xia, C; Advincula, RC. *Langmuir* 2003, 19, 4381.

[16] Fan, X; Xia, C; Advincula, RC. *Colloids and Surfaces A: Physicochem. Eng. Aspects* 2003, 219, 75.

[17] Fan, X; Xia, C; Advincula, RC. *Langmuir* 2005, 21, 2537.

[18] Vyazovkin, S; Dranca, I; Fan, X; Advincula, RC. *J. Phys. Chem, B* 2004, 108, 11672.

[19] Uthirakumar, P; Song, M.-K; Nah, C; Lee, Y.-S. *Eur. Polym. J*, 2005, 41, 211.

[20] Zhang, J; Yang, Y; Zhao, C; Zhao, H. *J. Polym. Sci., Part A: Polym. Chem*, 2007, 45, 5329.

[21] Yang, Y; Zhang, J; Liu, L; Li, C; Zhao, H. *J Polym. Sci., Part A: Polym. Chem*, 2007, 45, 5759.

[22] Zhang, J; Chen, K; Zhao, H. *J. Polym. Sci., Part A: Polym. Chem*, 2008, 46, 2632.
[23] Moet, A; Akelah, A. *Mater. Lett*, 1993, 18, 97.
[24] Moet, A; Akelah, A; Hiltner, A; Baer, E. *Proc. Mater. Res. Soc. San Francisco* 1994, April 2–8.
[25] Fu, X; Qutubuddin, S. *Mater. Lett*, 2000, 42, 12.
[26] Owusu-Adom, K; Guymon, CA. *Macromolecules* 2009, 42, 180.
[27] Böttcher, H; Hallensleben, ML; Nuß, S; Wurm, H; Bauer, J; Behrens, P. *J. Mater. Chem*, 2002, 12, 1351.
[28] Argoti, SD; Reeder, S; Zhao, H; Shipp, DA. *Polym. Prepr*, 2002, 43(2), 267.
[29] Zhao, H; Shipp, DA. *Chem. Mater*, 2003, 15, 2693.
[30] Shipp, DA; Zhao, H. *Polym. Prepr*, 2003, 44 (2), 249.
[31] Zhao, H; Argoti, SD; Shipp, DA; Farrell, BP; Shipp, DA. *J. Polym. Sci., Part A: Polym. Chem*, 2004, 42, 916.
[32] Zhao, H; Farrell, BP; Shipp, DA. *Polymer* 2004, 45, 4473.
[33] Yang, Y; Liu, L; Zhang, J; Li, C; Zhao, H. *Langmuir* 2007, 23, 2867.
[34] Di, J; Sogah, DY. *Macromolecules* 2006, 39(3), 1020.
[35] Weimer, MW; Chen, H; Giannelis, EP; Sogah, D.Y. *J. Am. Chem. Soc*, 1999, 121, 1615.
[36] Di, J; Sogah, DY. *Macromolecules* 2006, 39(15), 5052.
[37] Yang, Y; Wu, D; Li, C; Liu, L; Cheng, X; Zhao, H. *Polymer* 2006, 47, 7374.
[38] Zhou, Q; Fan, X; Xia, C; Mays, J; Advincula, R. *Chem. Mater*, 2001, 13(8), 2465.
[39] Fan, X; Zhou,Q; Xia, C; Cristofoli, W; Mays, J; Advincula, R. *Langmuir* 2002, 18, 4511.
[40] Usuki, A; Kojima, Y; Kawasumi, M; Okada, A; Fukushima, Y; Kurauchi, T; et al. *J. Mater. Res*, 1993, 8, 1179.
[41] Usuki, A; Kawasumi, M; Kojima, Y; Okada, A; Kurauchi, T; Kamigaito, O. *J. Mater. Res*, 1993, 8, 1174.
[42] Pantoustier, N; Alexandre, M; Degee, P; Calberg, C; Jerome, R; Henrist., C; et al. *e-Polymer* 2001, 9, 1.
[43] Pantoustier, N; Lepoittevin, B; Alexandre, M; Kubies, D; Calberg, C; Jerome, R; et al. *Polym. Eng. Sci*, 2002, 42, 1928.
[44] Messersmith, PB; Giannelis, EP. *J. Polym. Sci., Part A: Polym. Chem*, 1995, 33, 1047.
[45] Mariott, WR; Chen, E. *J. Am. Chem. Soc*, 2003, 125(51), 15726.
[46] Jin, Y.-H; Park, H.-J; Im, S.-S; Kwak, S.-Y; Kwak, S. *Macromol. Rapid Commun.* 2002, 23, 135.
[47] Shin, S.-YA; Simon, LC; Soares, JBP; Scholz, G. *Polymer* 2003, 44, 5317.
[48] Tudor, J; Willington, L; O'Hare, D; Royan, B. *Chem. Commun*, 1996, 17, 2031.
[49] Sun, T; Garces, JM. *Adv. Mater*, 2002, 14, 128.
[50] Wheeler, P; Mathias, L. *Macromol. Synth*, 2004, 13, 1.
[51] Wang, J; Wheeler, P; Baker, J; Mathias, L. *Polym. Prepr*, 2004, 45, 866.
[52] Wheeler, PA; Wang, J; Baker, J; Mathias, L. *J. Chem. Mater*, 2005, 17, 3013.
[53] Wheeler, PA; Wang, J; Mathias, L. *J. Chem. Mater*, 2006, 18, 3937.

[54] Herrera, NN; Letoffe, J.-M; Putaux, J.-L; David, L; Bourgeat-Lami, E; *Langmuir*, 2004, 20, 1564.

[55] Wu, Y; Zhang, J; Zhao, H. *J. Polym. Sci., Part A: Polym. Chem*, 2009, 3, 1535.

In: Polymer Nanocomposites:Advances in Filler Surface
Editor: Vikas Mittal
ISBN: 978-1-60876-125-8


*Chapter IV*

# Role of Monocationic and Bicationic Initiators on the Grafting of Polymer Chains 'from' the Clay Surface

***Periyayya Uthirakumar****

School of Semiconductor and Chemical Technology, Semiconductor Physics Research Center, Chonbuk National University, Jeonju, South Korea.

## Abstract

Grafting of polymer chains from the clay surface are made it successful with the help of cationic azo-initiators containing quaternary ammonium functional groups. Monocationic and bicationic initiators are designed for this purpose and a simple ion-exchange method is used to prepare initiator-bound clay layers. The interlayer *d*-spacing of the resulting initiator-bound clays are largely dependent on the size and their orientation of the organic initiators. The competence between the monocationic and bicationic initiators toward the surface functionalization upon ion exchange is discussed. The chain length and diameter of initiator molecules are facilities a higher *d* spacing in bicationic initiator-bound clay than the monocationic. The successful exfoliation of polymer clay nanocomposites resulted from the polymer grafting from the clay surface makes superiority over the other techniques. Hence, they offered a very high improvement in both thermal and mechanical properties, due to the high aspect ratio of individual clay layers that are uniformly dispersed throughout the polymer matrix.

**Keywords**: azo-initiators, monocationic and bicationic initiators, ion exchange, ammonium salts, d-spacing, exfoliation, nanocomposites.

---

* Corresponding author: uthirakumar@gmail.com

# 4.1. Introduction

Clay, in particular montmorillonite (MMT), a 2:1 phyllosilicate as illustrated in Figure 1, naturally occurs as stacks of platelets as in talc and mica but possesses different layer charge. The main source is from bentonite, a chemically altered volcanic ash that consists primarily of the clay mineral montmorillonite. The bentonite in its natural state can absorb up to seven times its weight in water, after treatment can absorb only 5 to 10 per cent of its weight in water, but 40 to 70 per cent in oil, grease, and other sparingly-soluble, hydrophobic chlorinated hydrocarbons. Chemically it is hydrated sodium calcium aluminium magnesium silicate hydroxide $(Na,Ca)_{0.33}(Al,Mg)_2(Si_4O_{10})(OH)_2 \cdot nH_2O$. Potassium, iron, and other cations are common substitutes, the exact ratio of cations varies with source. This leads to isomorphic substitution within layers (e.g. $Al^{3+}$ is replaced by $Fe^{2+}$ $Mg^{2+}$) and generates a negative charge exchange capacity (CEC) and for MMT. Figure 1 shows the ideal structure of MMT, 2:1 clay, means it has 2 tetrahedral sheets sandwiching a central octahedral sheet. The replacement of inorganic exchange cations with organic onium ions on the gallery surfaces of clays not only serves to match the clay surface polarity with the polarity of the polymer, but it also expands the clay galleries. Each platelet is less than 10 A°, thick, but over 200 times more in width.

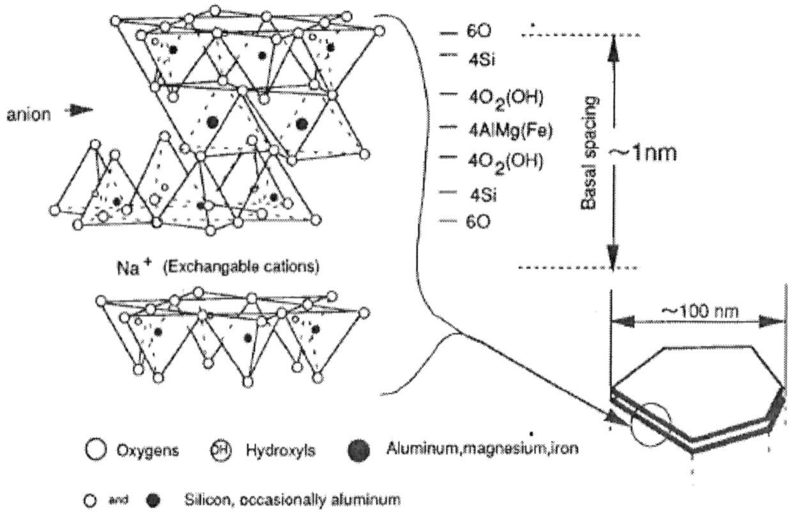


Figure 1. Idealized structure of MMT showing the 2:1 layered arrangement of a small section of two crystalline sheets of the material.

The gallery spacing separates each platelet. These spaces in MMT can be enhanced to 2-3 nm using quaternary ammonium salts. This facilitates the penetration of the gallery space by intercalation or exfoliation of either the polymer precursor or preformed polymer. Organoclays are manufactured by modifying clay surfaces with quaternary amines, normally a type of surfactant that contains a nitrogen ion [1-4]. The nitrogen end of the quaternary amine, the hydrophilic end, is positively charged and ion exchanges onto the clay platelet for sodium. In these silicates, regions between adjacent silicate layers contain highly mobile ions ($Na^+$, $Li^+$, $NO^{3+}$, etc.) that are readily extracted via exchange with other ions. If an ionic surfactant is used in this reaction, the result is an organosilicates, where the silicate layers were individually modified by the surfactant tail. It is therefore possible to produce a huge variety of organosilicates, based solely on different clays (variations in size and exchange capacity) and surfactants (variations in chemistry). With the proper choice of silicate and surfactant, a thermodynamically favorable nanocomposite structure can be formed; surfactant tails are displaced from the silicate layer surface, and are replaced by polymer chains that interact more strongly with either the silicate or the surfactants themselves. The result is a neutral surfactant with a solid base, which is the organoclay.

Towards the next step, intercalated or exfoliated polymer-clay nanocomposites can be achieved by using different type of functionalized organoclay particles. Polymer-clay nanocomposites are formed through the union of two very different materials with organic and mineral pedigrees. The hybrid compositions, however, exhibit large increases in tensile strength, modulus, and heat distortion temperature as compared with the pristine polymer [5-10]. The composites also have lower water sensitivity, reduced permeability to gases, and a similar thermal coefficient of expansion. All of these property improvements can be realized without a loss of clarity in the polymer. Further, it has been found that nanocomposites impart a level of flame retardance and UV resistance not present in the pure polymer. These improvements in performance properties at relatively low clay loading (typically 2 -10 wt%) have stimulated intensive research in both industry and academia over the past decade.

To explain the movement of clay particles upon the modification of organoclay followed by the polymerization is portrayed in Figure 2. A self-assembled two tetrahedral sheets are sandwiching with a central octahedral sheet is typical structure of pure MMT, as shown in Figure 2(a). As long as the extended organic molecular chains are penetrated in the self-assembled, well ordered gallery spaces, it is termed as intercalated structure as illustrated in Figure 2(b). The intercalated state is typically described as a situation where the silicate layers are expanded to accommodate some polymer molecules. However, the layers still retain a relatively small interlayer spacing, resulting in an ordered layered structure, analogous to the swelling of a multilayer stack. On the other hand, if the individual clay layers are no longer close enough to interact with adjacent layers it is termed as delaminated or exfoliated structure as explained in Figure 2(c). However, both of these hybrid structures can coexist. The enhancement in properties of polymer with modified MMT is mainly due to the clay which swells enormously in water and which in turn creates large surface area in which the polymer resides.

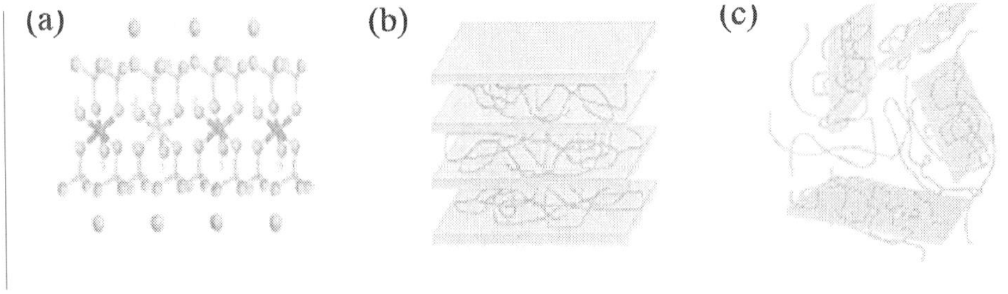


Figure 2. Structure of (a) pure MMT, (b) intercalated and (c) exfoliated polymer nanocomposites.

One of the most efficient methods to prepare the delaminated/exfoliated polymer-clay nanocomposites may be is to start polymerization in the clay galleries, using the ionically bound initiator [11-13], or monomer [14-16] anchored clay hybrid rather than the typical organoclays. The importance of in-situ surface initiated polymerization (SIP), as the polymer chain grows, the ordered silicate layers would be gradually pushed apart, ultimately exfoliating to discrete sheets. Here, at least some polymer chains must stay within the interlayer galleries at all times, preventing collapse of layers to an immiscible state. Additionally, the polymer-clay interface characteristics will change due to lower entanglement density, though higher interaction strength. This technique is amenable to both addition- and condensation polymerization. Although this method has been used throughout the history of polymer/layered silicate (PLS) nanocomposites preparation, it has limited application because of poor miscibility between organoclay and most monomers. Furthermore, a monomer suitable for intercalation is not always available for most of the technologically important polymers. If, instead, a polymerization initiator is used as the silicate compatibilizer, an extreme situation may develop. In this case, assuming that some monomer interaction can still occur and that no other source of initiation is added or present, every single polymer chain must form at the surface of a silicate layers, and must also remain end-tethered to that layer. Based on the above arguments, it can be classified as follows: They may also produce systems that are un-tethered (simple and emulsion in-situ polymerization), partially end-tethered (in-situ polymerization with monomer-modified silicate), or completely end-tethered (in-situ polymerization with initiator-modified silicate).

## 4.2. Significance of Initiator Bound Clay Layers for Polymer Grafting

As discussed earlier, naturally occurring MMT is hydrophilic, and as a result it is generally incompatible with most of organic materials (e.g. organic monomer and polymers). Therefore, it is necessary that they should be organically modified to improve compatibility with the most of polymers. Without compatibilization, such materials are less effectively particle-reinforced composites, called macrocomposites, here, the particles being large agglomerates (aggregation) of multi-layer silicate stacks. The property enhancement of polymer macrocomposites cannot reach to that of nanocomposites properties. In general, nano dispersed silicate particles showed more significant improvement than that of the micron size dispersed silicates layers in polymer matrix. In order to access the unique properties of nanocomposites, the dispersion must be significantly better throughout the matrix polymer.

The alkali cations can be exchanged out of the clay and be replaced with virtually any other cations. Various cation exchange reactions of MMT have been explained elsewhere in detail [3,4]. The process generally involves dispersing the clay in hot water in very low concentrations, typically ~1 wt%. To that, a second solution is added, where organic cations are present in a proper (calculated) amount. This proper amount of cation is calculated based on the cation exchange capacity (CEC) of the clay. The clay manufacturer often supplies this CEC values, but there are methods for determining the CEC value, when it is unknown

[17,18]. Popular organic cations are alkylammonium ions ("onium" ions). These molecules generally have one or more alkyl chains of 3-18 carbon members. The cationic ammonium moiety associates with the negative charges of the clay sheets, and the "greasy" chains extend away from the clay layers, generally leading to an increased gallery thickness. For this reason, clays exchanged with alkyl ammonium ions have spacing that are larger than the naturally occurring clay, and the interlayer spacing increase with increasing alkyl chain length, as summarized in Table 1.

**Table 1. List of the interlayer spacing of MMT (measured by X-ray diffraction) exchanged with a series of alkyl ammonium ions.**

| Cation | Interlayer d-spacing (nm) | Gallery thickness (nm) |
|---|---|---|
| $CH_3(CH_2)_3NH_3^+$ | 1.35 | 0.39 |
| $CH_3(CH_2)_7NH_3^+$ | 1.38 | 0.42 |
| $CH_3(CH_2)_9NH_3^+$ | 1.38 | 0.42 |
| $CH_3(CH_2)_{11}NH_3^+$ | 1.56 | 0.60 |
| $CH_3(CH_2)_{15}NH_3^+$ | 1.76 | 0.80 |
| $CH_3(CH_2)_{17}NH_3^+$ | 1.80 | 0.84 |

In place of carbon chain containing exchangeable ammonium cations, there are other specified molecules containing ammonium cations employed to study the dispersion of clay sheets effectively. For example, the layered silicates were treated with cationic monomers or organic initiators, to obtain the corresponding monomer-clay and initiator-clay hybrids, which can induce more efficient intra-gallery polymerization leading to the effective pushing of individual silicate platelets to far apart from each other, with no longer attractive interaction between the clay platelets.

The major significance of initiator bound clay layers are found to be the each and every single polymer chain must starts at the surface of a silicate layers, where the cationic initiator molecules are strongly attached at the clay surface, and must also remain end-tethered to that layer during free radical polymerization. Moreover, when the polymerization occurs from the clay surface, it generates more controlled grafts owing to the more controlled polymerization reaction due to the absence of external initiator. Two different types of cationic initiators (monocationic and bicationic) have been synthesized thoroughly studied for the suitability of polymer grafting by various research groups [11,13,19]. Among these initiator bound clay layers, a monocationic initiator attached to the clay surface is found to be superior over bicationic initiators in terms of low termination. However, the bicationic initiator also helps in reducing the amount of the solution polymer and if the surface is partially covered with the initiator, the possibility of the immediate termination of the radicals formed from the bicationic initiator can be expected to be substantially low. The advantage of in situ polymerization with the initiator bound clay that, as the bound polymer chains propagate, the ordered silicate layers can be gradually pushed apart and finally exfoliated in the polymer matrix. The extent of exfoliated clay platelets observed from the initiator bound clay system exhibit better dispersion in the polymer matrix than that of polymer clay nanocomposites obtained from the other techniques [20-28]

## 4.3. Types of Polymer Grafting

The surface initiated polymerization is carried out with the help of initiator attached clay surface with external monomers. SIP technique is used to graft the polymer chain through an in-situ free radical polymerization. Here, the polymerization initiators are inserted into clay galleries instead of reactive monomer cations or non reactive surfactants [15,16,29-31]. The surface-attached initiator not only renders the clay organophilic but also provides the initiation functionality for the subsequent in situ polymerization. Moreover, ionically bound oligomeric polymer chains polymerized "to" or "from" the clay surface can be specifically useful in terms of increasing basal plane spacing and hence exfoliation in the polymer matrices [11,13,31-44]. Also, the polydispersity in the chain length of these oligomers can help to reduce the autophobic dewetting at the interface, thus ensuring better mixing with the matrix [45]. Furthermore, as compared with the polymerizable surfactant route, the SIP strategy can presumably generate more end-tethered polymer chains with greater length, which provide more efficiency in exfoliation.

### 4.3.1. Polymer Grafting "to" the Clay Surface

This type of polymer grafting is generally achieved by functionalizing the "monomer" molecules with the exchangeable ammonium ion at one end. If the monomer containing an exchangeable ammonium ion, it can be easily exchanged with the sodium ions in the clay surface to make ionically bound monomers. The incorporating monomer molecules on the clay do not make any difficulties from the steric hindrance owing to their shorter chain lengths. Thus, the free radical polymerization starts with the available external monomer and initiator. A schematic diagram (Figure 3) demonstrates the dissimilarity between the grafting of polymer chains "to" and "from" the clay layers, on SIP technique.

### 4.3.2. Polymer Grating "from" the Clay Surface

The polymerization "from" the clay surface is almost same as that of polymerization "to" the clay surface, the only difference is, instead of attaching functionalized monomer, either mono or bifunctional cationic "initiator" molecules could be attached to the surface. Thus, the free radical polymerization can be initiates at the initiator bound clay layers followed by the grafting of polymer chains "from" the clay surface with the external monomers on the system.

The amount of clay surface coverage with the functionalized monomer or initiator is an important parameter, to achieve respective polymerization "to" or "from" the clay surface. The presence of initiator or monomer in higher amounts corresponding to cation-exchange capacity on one hand can be helpful in generating more radicals or polymerizing sites, but may also cause swift termination of the growing chains owing to the nearness of a large number of such sites to each other, therefore requiring a balance between the two factors along with the polymerization conditions.

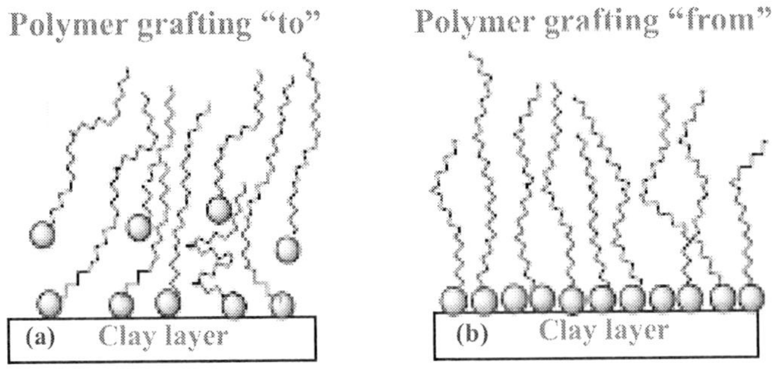


Figure 3. Drawing to distinguish the polymer grafting "to' or "from" the surface of the clay particles.

## 4.4. Types of Cationic Initiators for Polymer Grafting

Any organic molecule consists of quaternion ammonium salt functionality that can be useful for the modification of clay surface by cationic exchange mechanism. Hence, the well-known azo-initiators are the best choice to select as starting materials, as a core molecule, to prepare their own functional derivatives containing one or more quaternary ammonium salts at the chain end. Initiator bound clay layers can be prepared by the ion exchange reaction to obtain respective organophilic clays. Thus an in-situ polymerization is initiated by activation of the initiators that are ionically bound to the clay particle surfaces, leading to SIP process [11,13,19]. Owing to the high surface area of the montmorillonite platelets, these initiator species immobilized on the clay surface provide a suitable environment for the external monomer molecules to polymerize on the surface, leading to the generation of high-density polymer

Figure 4 illustrates the molecular framework of different types initiators are used for the SIP process. Designing and synthesizing cationic initiators molecules makes an initial step to prepare polymer clay nanocomposites. The synthetic scheme has to be design in such a way that, to incorporate as required number of exchangeable quaternary ammonium salts in the molecular chain. It can be one, two or more functional quaternary ammonium salts on the molecule chain. For example, if the cationic azo-initiator consists of only one quaternary ammonium salt at the end, named as "monocationic initiator", as shown in Figure 4(a). Only one quaternary ammonium bromide functional group is placed at one end of the chain with the free alkyl chain in other end. This type of initiator also named as "asymmetric type", due to the formation of non-identical halves upon cleavage, at the center of molecule. In contrary, when two quaternary ammonium functional groups are present in both ends then it will be "bicationic initiator", as shown in Figure 4b. In this case, both cleaved fractions are equivalent when the molecule is divided at the center; hence it is worth to call it as "symmetric type". On the other hand, if the cationic initiator has more than two quaternary ammonium functionalities then it may call it as "multicationic initiators" (Figure 4c). However, designing and synthesizing multifunctional initiators are extremely difficult and tedious; hence it became more expensive one for their commercial applications.

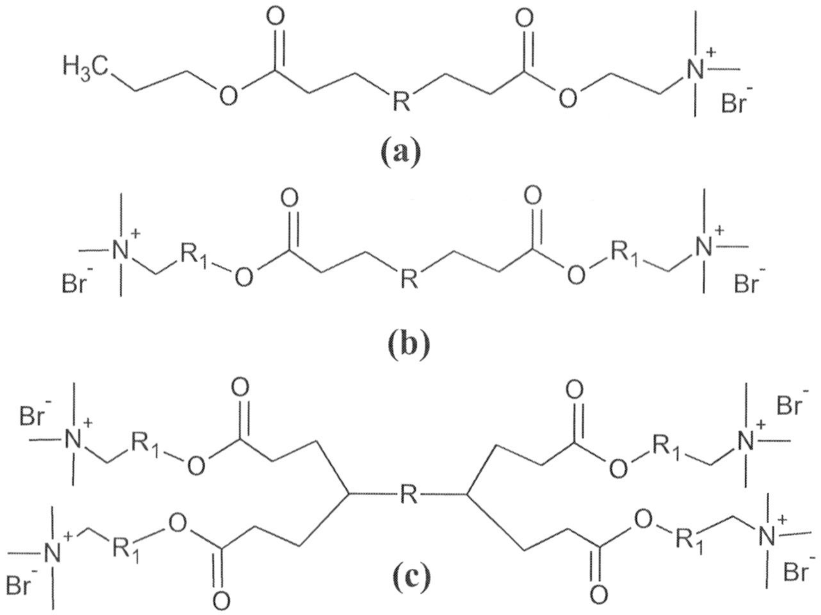


Figure 4. Molecular framework of different types of initiators: (a) monocationic, (b) bicationic and (c) multicationic initiators.

Till now, a very few research groups are actively participate in this area to prepare different types of cationic initiators. Also, hardly lesser number of initiators has been synthesized so for, to functionalize clay surface. Most of the these initiators are designed and synthesized from the commercially available azo-initiators, mainly from 4,4'-azobis(4-cyanovaleric acid) (ACVA) and 2,2'-azobis[2-methyl-N-(2-hydroxyethyl) propionamide], (AHPA). Fan et al proposed an esterification technique to synthesize the mono and bicationic initiators from the commercially available ACVA initiator [11]. The carboxylic acid is treated with bromoethanol by a well-known esterification method using DCC/DMAP [46]. In parallel, our group also effectively worked toward the design and synthesis of verity of bicationic initiators containing a choice of functional "onium" ion, from the commercially available AHPA initiator [13]. Similarly, Mittal et al also synthesized bicationic initiator having a longer alkyl bromoalcohol group to increase the length of carbon chain [19].

### 4.4.1. Synthesis of Monocationic Initiator (Asymmetric type)

A commercially available ACVA, azo initiator (11.2g), 2.96g of butanol, 0.06g of 4-(dimethylamino)-pyridine (DMAP), and 50ml of DMF were charged to a one-neck flask. The homogenous solution was cooled with an ice water bath. Then, 8.26g of 1,3-dicyclohexylcarbodiimide (DCC) was added in portions (Figure 5). The reaction mixture was allowed to react at 0°C for 5min and warmed to room temperature for 3 h. The urea was filtered as a byproduct, and the filtrate was poured into water, extracted with methylene chloride, washed by saturated sodium bicarbonate, and dried over magnesium sulfate. After solvent evaporation, the residue was solidified by pouring into cold hexanes. The solid (9.0g) was collected, dried, and used for the next step without further purifications. The crude product was reacted with 1 equivalent of bromoethanol under the same DCC/DMAP conditions to give the title compound. The product was purified by crystallization from methanol. 2.5g of white solid was obtained. The product was kept in a refrigerator in the dark. 1.5g of butyl 4-{2-[4- (2-bromoethoxy)-1-cyano-1-methyl-4-oxobutyl]diazenyl}-4-cyanopentanoate was stirred with excess of trimethylamine in anhydrous THF for two days. The precipitate was collected and washed thoroughly with THF. After drying under vacuum, 1.4g of the product was obtained.

### 4.4.2 Synthesis of Bicationic Initiator (Symmetric type)

Symmetrical type bicationic initiator also synthesized as like the synthetic scheme of monocationic initiator with a slight modification. Here, a similar way (Figure 5) is performed to get exchangeable quaternary ammonium bromide functional groups in both ends, without blocking the carboxylic acid by butanol group. The schematic representation for the preparation of bicationic initiator is presented in Figure 6.

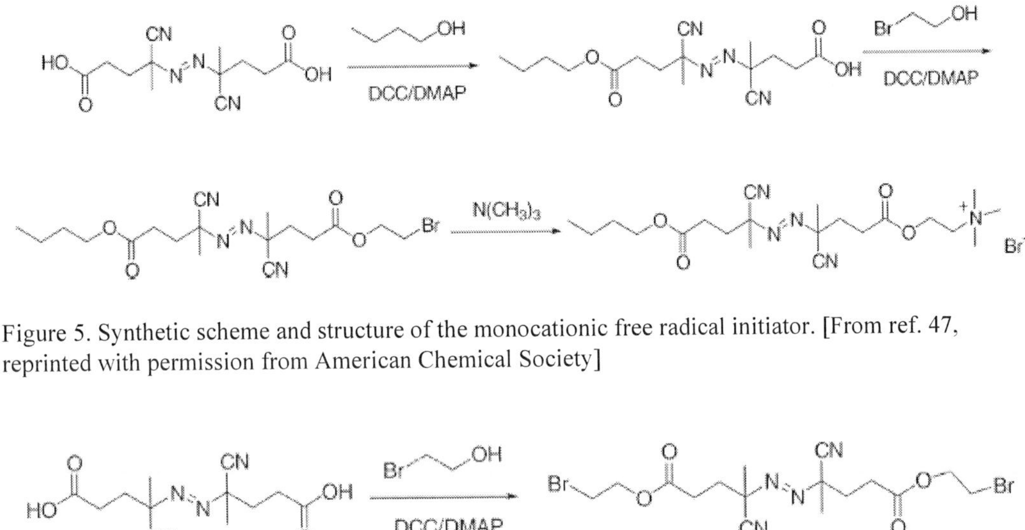


Figure 5. Synthetic scheme and structure of the monocationic free radical initiator. [From ref. 47, reprinted with permission from American Chemical Society]

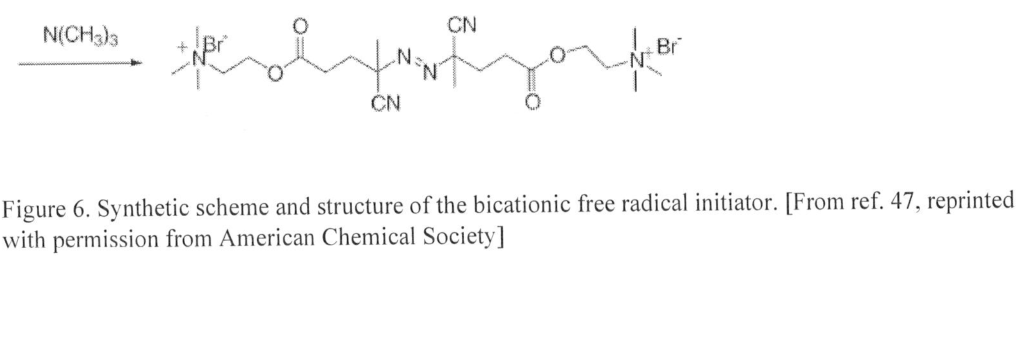


Figure 6. Synthetic scheme and structure of the bicationic free radical initiator. [From ref. 47, reprinted with permission from American Chemical Society]

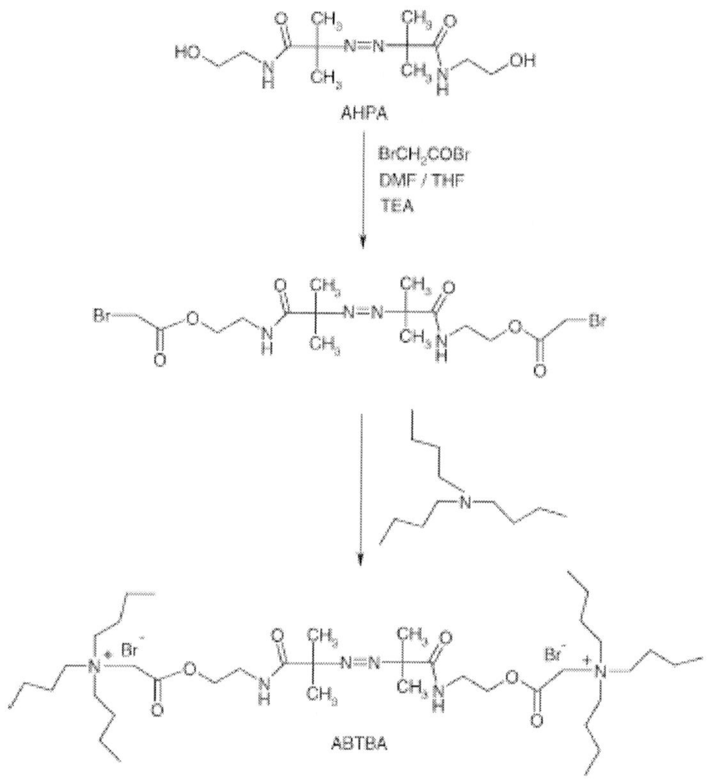


Figure 7. Synthetic scheme and structure of the bicationic free radical initiator (ABTEA). [From ref. 13, reprinted with permission from Elsevier]

We have designed a new type of symmetrical initiator from the commercially available AHPA initiator, having primary alcohol groups in both ends. A simple esterification technique has been followed with the help of available reactive primary alcoholic functional groups in the AHPA, a starting raw material, using a bromoacetyl bromide (acyl halide), followed by the subsequent quaternization reaction with the different types of tertiary amines Figure. 7 illustrates the schematic diagram explains the synthetic procedure to obtain a 2,2'-azobis{2-methyl-N-[2-acetoxy-(2-N,N,N-tributylammonium bromide)ethyl] propionamide}, (ABTBA), where the tributylamine is used as a tertiary amine to obtain a quaternary ammonium salt.

In details, the commercially available starting raw material AHPA (1.0 g, 3.45 mmol) was dissolved in 10 ml of DMF and 33 ml of THF, followed by the addition of bromoacetyl bromide (1.66 g, 8.22 mmol) as shown in Figure 7. The homogeneous solution was stirred for 20 min at 25 °C, and then cooled to 0°C in an ice bath. Triethylamine (0.85 g, 8.41 mmol) was added dropwise. The reaction mixture was allowed to react at 0-5°C for 2 h. The triethylammonium bromide, formed during the reaction, was removed by filtration, and the resulting solution subjected to the reaction with excess tributylamine, for 13 h, at about 10°C. The reaction mixture was extracted several times with ether. The product was dried under reduced pressure, at room temperature, to get pure ABTBA, as shown in Figure 7.

Similarly, other tertiary amines such as pyridine (Py) and triethyl amine (TEA) are used to make different types of quaternary ammonium functional groups, as represented as 2,2'-

azobis{2-methyl-N-[2-acetoxy-(2-pyridinium bromide)ethyl] propionamide}, (ABPy) and 2,2'-azobis{2-methyl-N-[2-acetoxy-(2-N,N,N-triethylammonium bromide)ethyl] propionamide}, (ABTEA), respectively. Figure 8 depicts the three different symmetrical bicationic azo-initiators consist of different types of quaternary ammonium salts in both ends of bicationic initiators.

Mittal et al. proposed a new way to lengthen the molecular chain by incorporating a longer alkyl chain bromoalcohol during preparation of bicationic initiator synthesis, as shown in Figure 9. The amounts of 5.6 g (20 mmol) of ACVA, 15 g (60 mmol) of 1-bromo-11-undecanol, 0.06 g of DMAP, and 30 mL of dimethylformamide (DMF) were placed in a round-bottom flask. The contents were allowed to dissolve fully and the homogeneous solution was cooled in an ice water bath followed by the addition of 10.30 g (50 mmol) of DCC. DCC was allowed to dissolve fully and the reaction mixture was allowed to further react for 5 min at 0 °C followed by warming to room temperature.

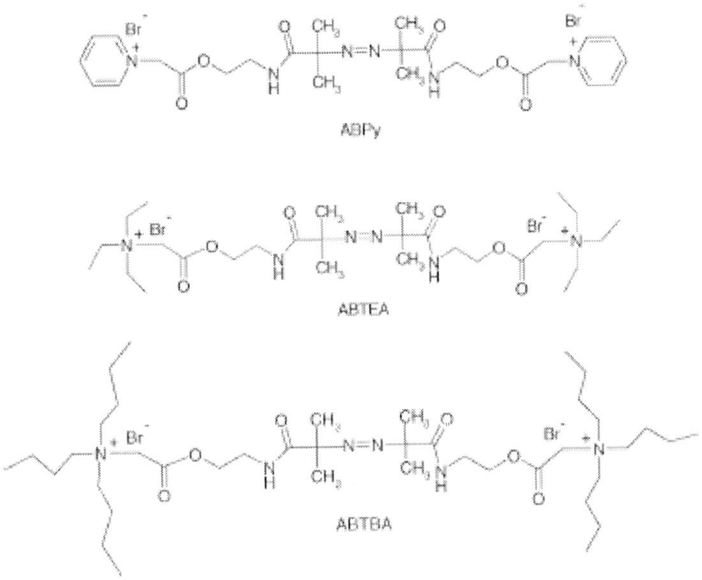


Figure 8. Chemical structures of ABPy, ABTEA and ABTBA initiators. [From ref. 13, reprinted with permission from Elsevier]

After running the reaction at room temperature for 24 h, the urea formed during the reaction was filtered and the filtrate was poured in water and extracted with methylene chloride followed by washing with saturated sodium bicarbonate and drying over sodium sulfate. The solvent was then evaporated, and cold methanol was added to the residue which led to the generation of the crystals of the dibromo product formed, which were subsequently filtered.

In the subsequent quaternization step, these crystals (4.2 g, 5.65 mmol) were dissolved in 25 mL of THF followed by slow addition of an excess of trimethylamine (16.95 mmol) and the reaction was kept under stirring at room temperature for 48 h.

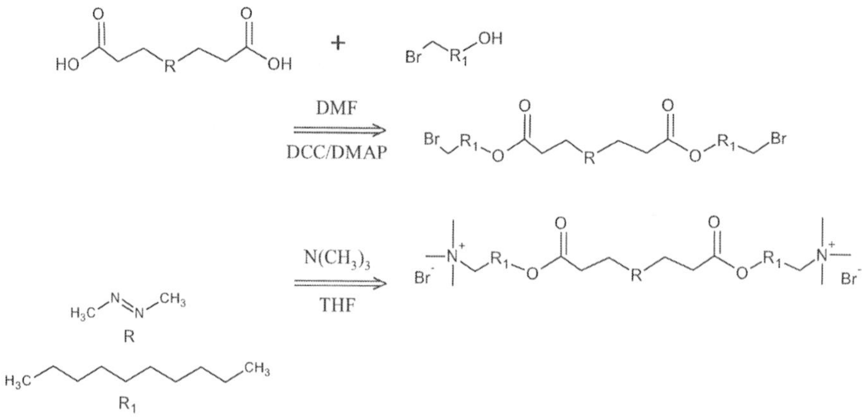


Figure 9. Synthetic scheme and structure of the bicationic free radical initiator.

THF was then evaporated and the crystals of the bicationic salt were precipitated by the addition of ether. Repeated extraction with diethyl ether led to an overall yield of 80%.

## 4.5. Cationic Initiator-Bound Clay Layers

To make effective SIP process, free sodium ions in the clay surface has to be completely replaced by the cationic initiator molecules, following the free radical polymerization leads the exfoliated polymer nanocomposites. Before going into the deep, there may be an intention to learn how to replace all of sodium ions inside the clay galleries and their confirmation of the ionically bound initiators on the clay surface. In order to obtain the initiator-bound (surface-attached) initiator, a simple cation exchange process has been preformed. In general, the positively charged initiator can be exchanged with the sodium ion in the clay galleries is demonstrated by the following reaction:

$$Na + clay - + Initiator + Br \Longrightarrow - Initiator + clay - + Na + Br -$$

Assuming that the available sodium ions are perfectly replaced by the initiator cations, followed by the thorough washing to remove excess initiators, there will not be free sodium ions remaining in the clay layers. The total exchange of sodium ion by the initiator cation has been confirmed by the various techniques, such as X-ray photoelectron spectroscopy (XPS), Infrared spectroscopy (FT-IR) and thermo gravimetric analysis (TGA), etc. Figure 10a shows the XPS scan of pure clay and intercalated clay samples. In the pure clay spectrum, it is obviously found the characteristic peaks of O, Na, Si, Al, and Mg in the pure clay. However, upon cationic exchange with initiators, the Na signal became very weak and almost disappeared on the spectra of Figure 10(b-c). Besides the appearance of new C 1s and N 1s peaks are clearly demonstrates that the presence of organic initiators in the spectra (b) and (c), in Figure 10A. This confirms the perfect exchange of sodium ions in the clay sheets by the cationic initiators.

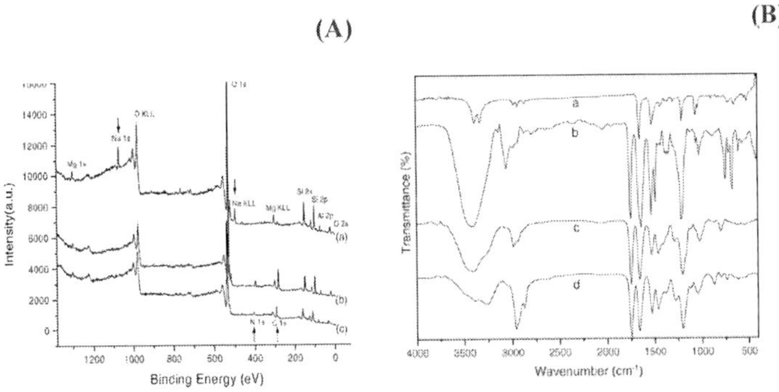


Figure 10. (A) XPS spectra of (a) pure clay, (b) monocationic initiator and (c) bicationic initiator attached with the clay surface. (B) FT-IR spectra of bicationic initiator-bound clay systems containing (a) AHPA, (b) ABPy, (c) ABTEA and (d) ABTBA. [From ref. 13,47, reprinted with permission from American Chemical Society and Elsevier]

FT-IR spectroscopy can be another alternative method to confirm the total exchange of cationic initiators with the sodium ions. The FT-IR spectra of the $Na^+$-MMT and the surface attached initiator are shown in Figure 10B. The $Na^+$-MMT spectrum shows characteristic O–H stretching, interlayer water molecules deformation, Si–O and Al–O peaks, at 3627, 1635, 1091/1039 and 519/466 $cm^{-1}$, respectively. The characteristic peaks of the respective functional groups of the initiators are clearly seen. For example, absorption peaks at 3450-3350, 1550-1480, 1750, 2800-2900, and 3100 $cm^{-1}$ are designated to be secondary amide N-H, C-N stretching, ester C-O, aliphatic C-H, and aromatic C-H stretching characteristic, respectively. Yet another effective method is TGA technique, where the decomposition trend of both components in the initiator bound clay layers can be noticed, upon thermal treatment. Figure 11 shows the TGA curves of the initiators attached clay surface show two distinctive degradation stages, in addition to the weight loss at around 100°C, due to free water molecules. The first major decomposition starting at 220°C, corresponding to the degradation of the azo group in the organic initiators and leading to the evolution of N2 gas. The second degradation, at around 420°C, is due to the decomposition of the rest of the organic initiator molecules. Based on this information, one can easily substantiates the presence of both counterparts in each initiator-bound clay samples.

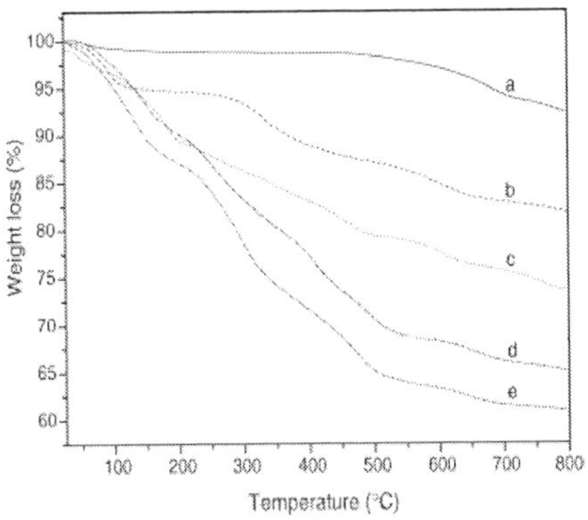


Figure 11. TGA curves of (a) pure clay and bicationic initiator-bound clay systems containing (b) AIBA, (c) ABPy, (d) ABTEA and (e) ABTBA. [From ref. 13, reprinted with permission from Elsevier]

### 4.5.1. Preparation of AHPA Derivative Initiator Bound Clay Layers

The AHPA derivative initiator-bound clay layers are prepared by cationic exchange reaction between the $Na^+$ ion in the clay galleries and the quaternary ammonium or pyridinium part of the initiators. $Na^+$-MMT (3.0 g) was suspended in 200 ml of deionized water at 70°C, for 1 h. The suspension was slowly cooled back to room temperature, and added to an aqueous solution (200 ml), containing 1.5 mol equivalents of each azo-initiator, for example, 3.92 g is used in case of ABTEA. The mixture was stirred for 20 h at room temperature, to ensure complete exchange. The white precipitate formed during the exchange reaction was filtered and re-dispersed in water. This procedure was repeated until the chloride ions and unintercalated initiators were removed. The resulting solid was washed with methanol and dried under vacuum in the presence of anhydrous calcium chloride at 35°C, for 48 h.

Figure 12 shows the XRD patterns of the Na+-MMT and the various organic initiator-MMT hybrids. 2,2'- Azobis(isobutyramidine) dihydratochloride (AIBA), a similar type of azo-initiator, commercially available, is used for comparison. Na+-MMT has an interlayer d-spacing of 1.20 nm, as calculated from the Bragg's equation $n\lambda = 2d \sin \theta$, where $\lambda = 0.1543$ nm. The interlayer d-spacing of the initiator-bound clay is increased from the base value (1.20 nm) to 1.52 (AIBA-MMT), 1.58 (ABPy-MMT), 1.86 (ABTEA-MMT) and 2.06 nm (ABTBA-MMT).

For a given initiator-bound clay system, the interlayer d-spacing probably depends on the orientation and size, or chain length of the initiator molecules, as long as the concentrations of the intercalated initiators are the same. As explained earlier, based on the TGA data (Figure. 11), the AIBA-MMT, ABTEA-MMT and ABTBA-MMT hybrids are prepared via complete exchange of the Na+ ions in the Na+-MMT, with the respective cationic initiators. Thus, it should be reasonable to assume that the concentrations of the intercalated initiators in the clay galleries are the same for the three different initiator-bound clay systems.

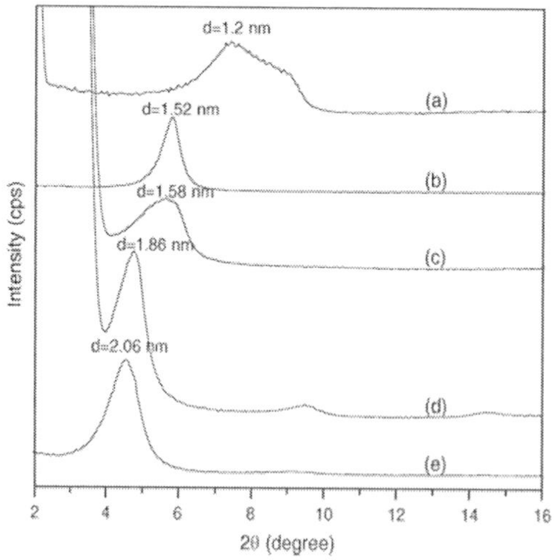


Figure 12. XRD patterns of (a) Na+-MMT, (b) AIBA-MMT, (c) ABPy-MMT, (d) ABTEA-MMT and (e) ABTBA-MMT. [From ref. 13, reprinted with permission from Elsevier]

### 4.5.2. Preparation of ACVA Derivative Initiator Bound Clay Layers

Dispersion of MMT clay layers by vigorous stirring on overnight standing will helps to obtain better insertion of initiators inside the clay space by effect ion exchange reaction [47,48]. MMT clay (1.3 g) was dispersed in 500 mL of deionized water, and the dispersion was stirred vigorously overnight at room temperature. The clay suspension was then ultrasonicated while stirring for 24 h, and a yellowish, homogeneous dispersion was obtained. Both mono and bicationic initiators were dissolved in deionized water to make 1.25 wt% aqueous solutions. The exact weight (the weight of the initiator needed to fully replace the Na counter ion of the clay used) of the two initiators was calculated. Accordingly, an excess amount (about 23% more) of initiator solution was slowly poured into the clay dispersion while it was being stirred and ultrasonicated. The mixture was stirred and ultrasonicated for another 24 h. The resulting white precipitate was separated by filtration and then re-dissolved in deionized water. This filtration–dissolution process was repeated 10 times in order to remove the free, unattached initiator.

XRD patterns of the pure clay, monocationic and bicationic initiator-intercalated clay samples are shown in Figure 13. The d spacing values of the samples are calculated and shown beside each peak. The basal spacing of pure $Na^+$-MMT is 1.16 nm, which is in accordance with data from other sources [49]. The XRD patterns of the intercalated clays indicate the successful insertion of the initiator molecules into the galleries of the silicate platelets since both intercalated clay samples gave increased d-spacing values. Also, the sharper shape and the higher diffraction intensities of the peaks after intercalation provide more evidence in support of a better-ordered swollen structure than that of the original clay. However, the d-spacing of bicationic initiator is much smaller than that of the monocationic initiator bound clay. The difference in their d-spacing is purely depending on the molecular orientation inside the clay galleries that will affect the interlayer distance between the clay platelets.

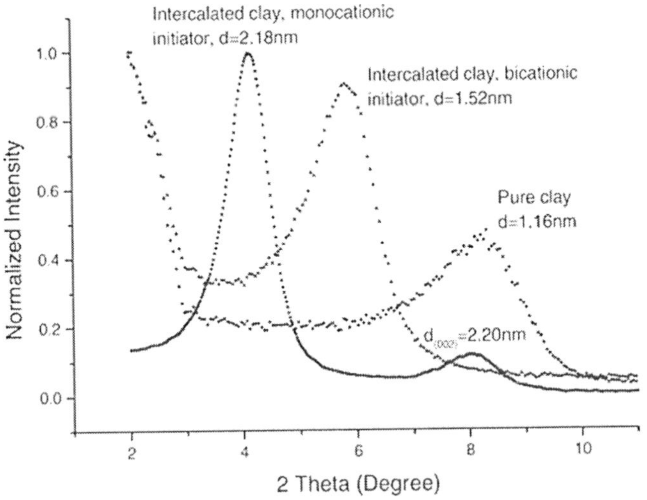


Figure 13. X-ray powder diffraction patterns of pure clay and two intercalated clay samples. [From ref. 47, reprinted with permission from American Chemical Society]

### 4.5.3. Partial Initiator Bound Clay Surface with Other Alkylammonium Surfactant

Another interesting idea is utilized to modify the clay surface with the partial insertion of bicationic initiator along with the other alkyl ammonium chains in amounts corresponding to varying percentages of clay-cation-exchange capacity [19]. A normal cation exchange reaction is performed with bicationic initiator used in a quantity corresponding to 30% of the CEC and dihexadecyldimethylammonium bromide corresponding to 70% of the CEC, and vice versa, leading to the d-spacing is about 2.2 and 1.9 nm, respectively.

## 4.6. Possible Orientations of Initiator Molecules on the Clay Surface

Figure 14A illustrates the possible orientations of anchored azo-initiators inside the clay galleries. The interlayer d-spacing is expected to be increase in the direction of the down-arrow. Being all of our azo-initiators are having two ammonium groups in both the ends, there may be lots of possibility of surface orientations.

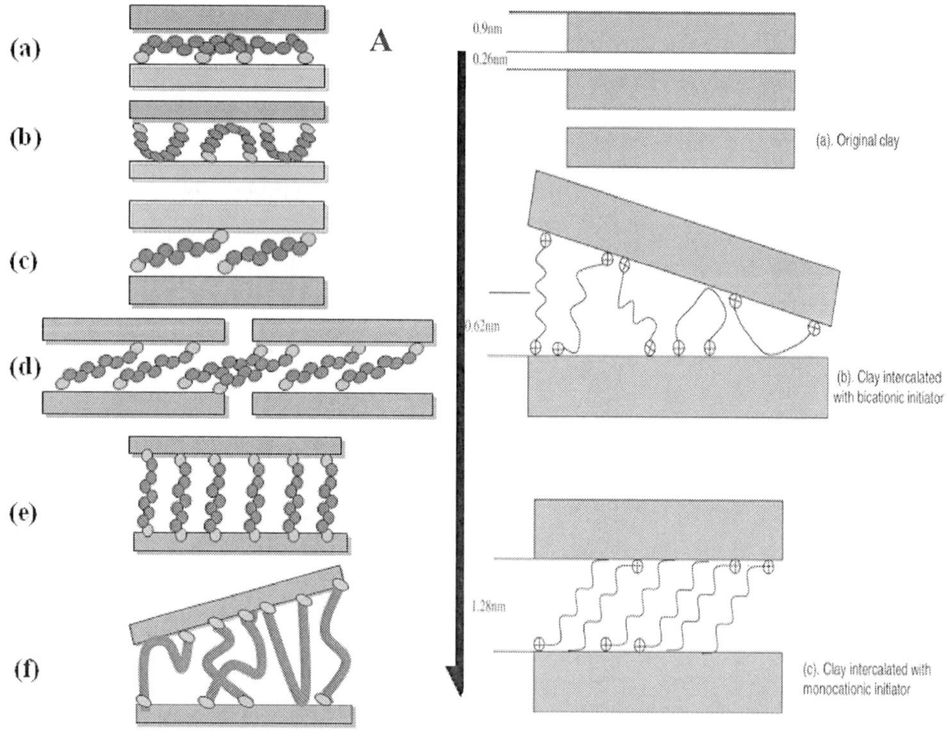


Figure 14. (A) Possible orientation of bicationic initiators inside the clay galleries in an order of increasing the interlayer distance (top to bottom) and (B) Schematic diagrams of the intercalation: (a) original clay, (b) clay intercalated with bicationic initiator, and (c) clay intercalated with monocationic initiator. [From ref. 47, reprinted with permission from American Chemical Society]

Based on imaginations the following are the some of the surface treated orientations. One way could be the chance to exchange the both ions in the same silicate sheets (a), or the same with alternative manner as mentioned in (b). On the one hand, each end of ammonium ion may exchange with adjacent silicates sheets (c) effectively, and also another end may exchange with yet another silicate sheet in the sidewise, as shown in (d). However, in all above cases, most of the initiators molecules were lied down with the silicate sheet in parallel fashion, and there may be chances to standing up position too. On the other hand, the image (e) explains that all initiator molecules are standing up (straight) in a regular fashion. Overall, these kinds of different possible orientations of organic initiator molecules result in a random distribution of interlayer d-spacing, leading to the broader X-ray diffraction peaks. This result demonstrates that the layered framework of inorganic clay can accommodate initiator molecules of various functionalities with better long-range periodicity.

The difference in molecular orientation of initiator moieties inside the clay layers between the monocationic and bicationic initiators is also investigated. The molecular chain length of the initiator moieties will adjust the interlayer spacing between the clay platelets. Based on the XRD data, the interlayer distance after exchange with monocationic and bicationic initiators is found to be 2.18 and 1.52 nm, respectively. However, the d-spacing values seemed to be inconsistent with the steric sizes of the two different initiators. The d-spacing of bicationic intercalated clay (1.52 nm) is much smaller than that of the monocationic intercalated clay although their molecular dimensions are about the same (both chain length values are 2.20 nm, estimated by Chem 3D software). It has been believed that the reason for this observation lies in the presence of charged groups on both ends of the bicationic initiator molecule. Thus, there are mainly two intercalation possibilities: (1) the two cationic endgroups interact electrostatically with two different but neighboring platelets' surfaces, or (2) they interact on the same side of a single clay particle. The combination of these two possibilities makes the intercalated structure less spatially ordered which could account for the significantly broadened reflection for this sample as compared, as shown in Figure 14B. However, d-spacing value gets improved further by incorporating a longer chain length and diameters of the bicationic initiators.

We put another step forward to investigate the effect of chain length and diameter of initiator molecules on the interlayer spacing enlargement, after exchange with the clay surfaces. Thus, a short to lengthy carbon chains containing exchangeable ammonium cations has been synthesized to achieve higher interlayer d-spacing of the silicate sheets, which could create an enough space for the efficient intercalation of monomer molecules to achieve perfect exfoliated polymer nanocomposites. The theoretical values of chain length and diameter of the synthesized initiators molecules are calculated using Hyperchem software. The values were estimated after MM+ energy minimization technique. The molecular views obtained after the energy minimization steps are shown in Figure 15. The corresponding results were listed in Table 2, the chain length of commercially available AIBA initiator was 0.8 nm, but those of the synthesized initiators were calculated to be 1.41, 2.62 and 3.04 nm for ABPy, ABTEA and ABTBA, respectively. Coherently, the initiator diameters also increased in the series from AIBA to ABTBA, as mentioned in the Table 2. These results confirm that the chain length and diameter of bicationic initiator molecules help to enlarge the interlayer spacing upon intercalation.

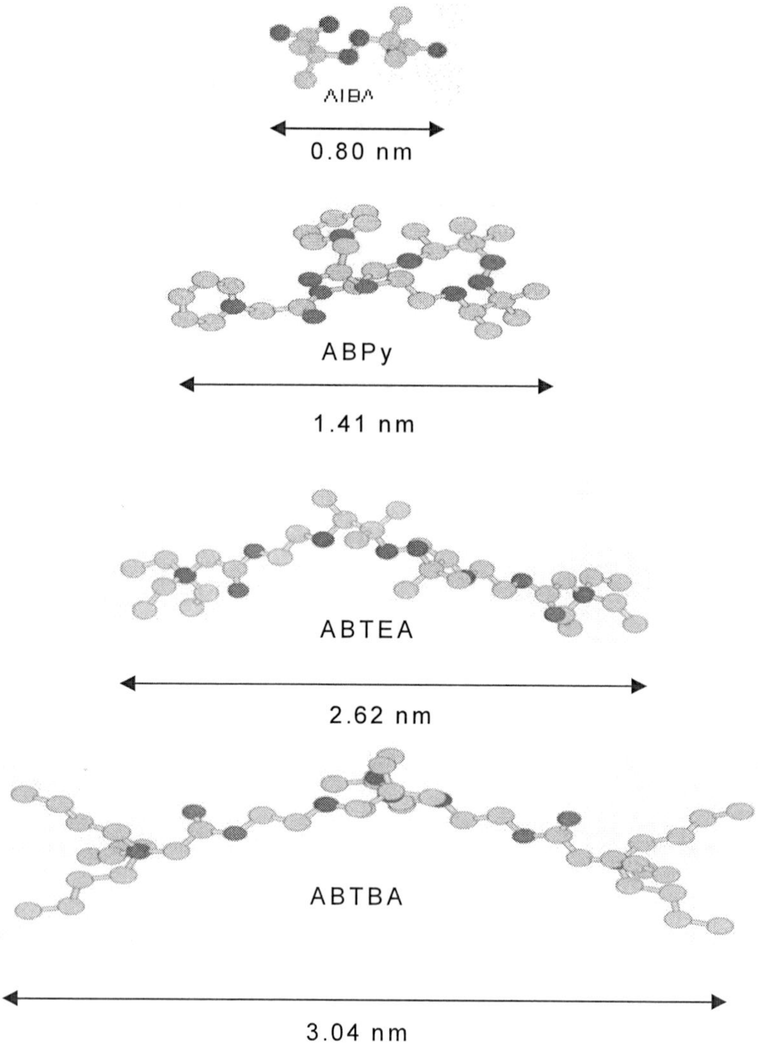


Figure 15. The molecular views of bicationic initiators obtained after the MM+ energy minimizing technique. [From Ref. 13, reprinted with permission from Elsevier]

**Table 2. Molecular dimensions of the initiators were calculated based on the HyperChem software.**

| Sample code | Chain length (nm) | Chain diameter (nm) | Interlayer d-spacing* (nm) |
|---|---|---|---|
| AIBA | 0.80 | 0.49 | 1.52 |
| ABPy | 1.41 | 0.87 | 1.58 |
| ABTEA | 2.62 | 0.86 | 1.86 |
| ABTBA | 3.04 | 1.11 | 2.06 |

*Interlayer d-spacing calculated from the XRD data

## 4.7. Role of Bicationic Initiators for the Polymer Grafting

Towards the modification the hydrophilic clay into hydrophobic clay surface, various types of organic molecules consists of quaternary ammonium salts are used to exchange effectively. Thus, in-situ polymerization is initiated by activation of the initiators that are ionically bound to the clay particle surfaces, through a SIP process. As we discussed earlier that, the advantage of SIP process is where the polymer chain grows after surface initiation, the ordered silicate layers could be gradually pushed apart, ultimately exfoliating to discrete laths, resulting in a well-dispersed structure of the final product. Also, theoretically, if all initiators are tethered to clay surfaces, a higher efficiency of inter-gallery polymerization is expected compared to that of free, unattached initiators

Exfoliated PS/MMT nanocomposites are prepared successfully from the in-situ free radical polymerization techniques. Here, bulk polymerization technique is used to prepare PS/MMT nanocomposites using 1, 3 and 5 wt% ABTBA initiator-bound clay are designated PS/MMT-1, PS/MMT-3 and PS/MMT-5, respectively. In Figure 16, the diffraction peak of the ABTBA initiator-bound clay appears at the 2θ value, 4.3°, which corresponded to an interlayer d-spacing of 2.06 nm based on Bragg's equation. The interlayer d-spacing of the ABTBA initiator-bound clay is higher than that of pure Na-MMT (1.2 nm), due to the intercalated azo-initiator molecules in the clay galleries. The PS/MMT-1 and PS/MMT-3 nanocomposites showed no noticeable diffraction peak at the 2θ value, from 1 to 10°. This result indicated that the clay platelets are exfoliated and dispersed in the PS matrix, where the interlayer distances are greater than 6 nm. On the other hand, the PS/MMT-5 showed a weak and broad peak at the 2θ value, of 6.2°, as shown in the inset to Figure 16.

The XRD patterns of the inset were recorded for the same samples in the lower 2θ region, by varying the divergence slit to distinguish more clearly between the diffraction patterns. This result indicated that the average interlayer d-spacing of the clay platelets in the polymer matrix was 1.42 nm, which lay between the interlayer d-spacing of the Na-MMT and ABTBA-MMT. To resolve this issue, a new concept of viscosity competition between the intra- and extra-gallery phases, during the in situ polymerization is considered. The exfoliation was only possible when the intra-gallery polymerization was at least equal to, or faster, than that of the extra-gallery polymerization [50].

The exfoliation trend of the clay sheets in the PS/ MMT nanocomposites is also confirmed by TEM, as shown in Figure 16. The dark lines correspond to the silicate platelets in the polymer matrix. The micron-sized silicates are not observed in the PS/MMT-1 image, but instead, the dispersed silicate nanoplatelets of the large aspect ratios had thickness less than 20 nm. The silicate nanoplatelets are oriented somewhat parallel to one another, but largely separated from each other, so they could not interact any more. A similar degree of exfoliation and dispersion is observed in the PS/MMT-3, but the thicknesses of the dispersed silicate nanoplatelets are slightly higher than 20 nm. However, the PS/ MMT-5 exhibited a somewhat different morphology of its silicate layers from the others. There are some silicate aggregates, along with fewer exfoliated nanoplatelets. This result is consistent with the XRD data, as described above.

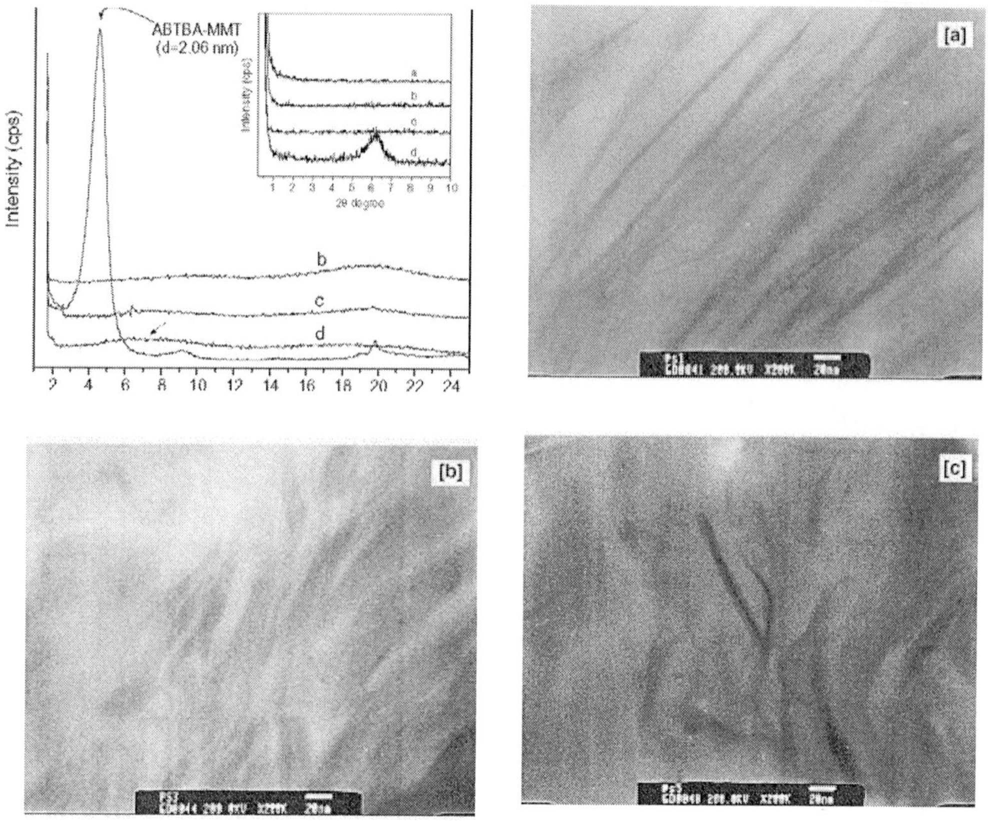


Figure 16. XRD and TEM images of (a) PS/MMT-1, (b) PS/MMT-3 and (c) PS/MMT-5. The scale bar represents 20 nm. [From ref. 41, reprinted with permission from Elsevier]

Mittal et al carried out the solution polymerization with the bicationic initiator along with other alkyl ammonium ions in different proportion to achieve the exfoliation of lauryl methacrylate [19]. A bicationic initiator carrying longer alkyl chains between the initiator moiety and the ammonium head groups has been synthesized. Cation exchange on the clay surface with bicationic initiator added in a quantity corresponding to 30% of the CEC and dihexadecyldimethylammonium bromide corresponding to 70% of the CEC led to an exchange of 91% on the clay surface. Based on the X-ray data, the clay with initiator exchanged with an amount corresponding to 30% of CEC showed at a d-spacing of 2.2 nm. When reacted at 60 °C for 72 h under nonliving conditions, the d-spacing increased to 2.4 nm. At the same time, when reacted at 120 °C using the living polymerization conditions, the diffraction angle is shifted to lower degree corresponding to a d-spacing of 3.4 nm. Similarly, the organically modified clay with initiator present in an amount of 70% of CEC has the d-spacing of 1.9 nm, which remained unchanged when reacted at 60°C for 72 h under nonliving conditions as is evident from a minor increase in organic weight loss. But when reacted at 120°C under the living conditions, the d-spacing is increased to 3.8 nm. It indicates that apart from effective control over the termination reactions by using the living agent, the higher extent of coverage of the surface with initiator is also beneficial in further enhancing the basal plane spacing of the clay platelets. Thus, it confirmed the successful generation of the polymethacrylate brushes grafted from the clay surface.

# 4.8. Role of Monocationic Initiators for Polymer Grafting

Figure 17 depicts the PMMA-MMT nanocomposites prepared with the help of intercalated monocationic initiator in bulk, solution and suspension method via SIP [48]. For bulk and suspension SIP, 0.071 g of intercalated clay was added to 10 mL of MMA monomer (pure clay loading 0.5 wt% based on MMA). For solution SIP, the same amount of intercalated clay was added to a mixture of 10 mL of MMA monomer and 10 mL of tetrahydrofuran. All three dispersions were ultrasonicated long enough (usually 6-10 h with active ~30 min cycles and dormant ~5 min intervals) until a homogeneous system was observed. For suspension SIP, the intercalated clay-monomer dispersion was added dropwise to 50 mL of deionized water with vigorous stirring. The three polymerization systems were purged with N2 flow for 1 h, and the SIPs were carried out at 60°C for 72 h under N2 protection. Moreover, a pure PMMA sample was synthesized as a reference by solution polymerization using recrystallized AIBN initiator under the same conditions.

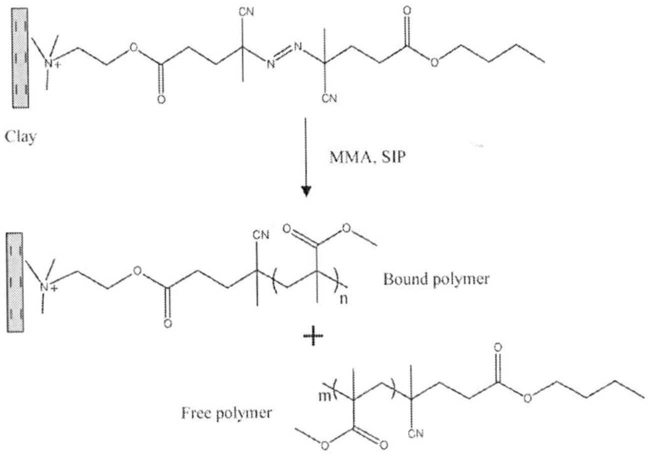


Figure 17. Schematics of the synthesis of PMMA–clay nanocomposites by free radical SIP from clay surfaces. [From ref. 48, reprinted with permission from American Chemical Society]

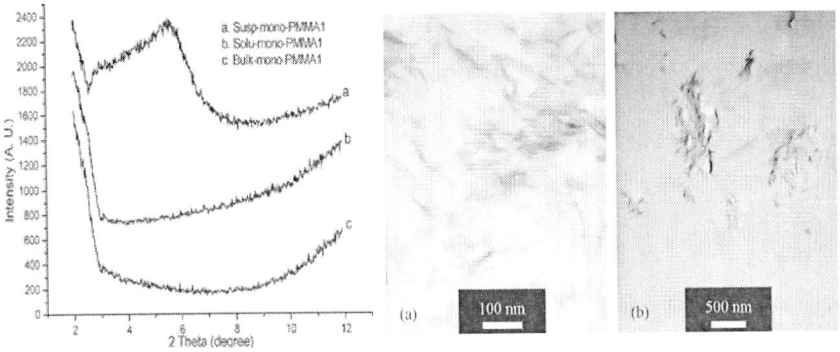


Figure 18. XRD diffractograms of three nanocomposite samples and TEM images of (a) Solu-PMMA1 and (b) Susp-PMMA1. [From ref. 48, reprinted with permission from American Chemical Society]

The XRD results show that an exfoliated structure was achieved by solution and bulk SIPs, as their diffractograms in Figure 18 show a featureless pattern. This is confirmed by TEM where an image of disordered clay platelets being well dispersed in polymer matrix was observed in Figure 18(a). Again, the homogeneity of the initiator–MMA–clay system during SIP reactions should have contributed to the well-dispersed structure. More importantly, it has been believed that the dense tethering of the long polymer chains is the critical factor for exfoliation. Tethered polymer chains can promote and preserve the exfoliated morphology by pushing and keeping the silicate layers apart. This key concept has been confirmed by a considerable amount of studies including polymer–clay nanocomposites via the free radical polymerization of other vinyl monomers [15,16,51] and even other polymerization mechanisms [4,52,53].

In the case of suspension SIP, compared to those of the pure and intercalated clay, the broad peak suggests that the clay particles have a lesser degree of order but with still moderate short-range periodicities, that is, partially exfoliated structure. This is verified by the TEM observation shown in Figure 18(b) which exhibits clay aggregates of various sizes in sample Susp-PMMA1. For suspension polymerization, it is well-known that silicate clays such as montmorillonite can be used as the dispersant to prevent monomer droplets from re-agglomerating. In the case of our suspension SIP, the initiator-attached clay not only serves as the suspension stabilizer but also activates the polymerization. Through a similar approach of suspension polymerization of MMA, Huang et al has obtained intercalated and exfoliated structures by alkylammonium- and AIBA-modified clays, respectively [54].

## 4.9. Monocationic vs Bicationic Initiators

In order to compare the extent of exfoliation between the monocationic and bicationic initiator attached clay surface, a solution polymerization has been carry out via SIP. It has been found that efficient clay nanoparticle dispersion and monomer diffusion are crucial factors in performing a successful SIP and in achieving complete exfoliated clay structures. The X-ray diffractograms of the two SIP products are shown in Figure 19. Sample bi-PS-M-2 shows a small peak at $2\theta = 5.9°$, which is about the same as the peak position of the corresponding intercalated clay. This result indicates that this product still contains intercalated clay structure.

On the contrary, there is no peak on the XRD trace of sample mono-PS-M-2, indicating a completely exfoliated structure was achieved from the monocationic initiator attached on the clay surface.

In general, for the bicationic initiator attached clay system, adjacent clay layers would be gradually pushed apart, once the two immobilized free radicals were simultaneously generated. As a result, the intercalated clay stacks would be totally delaminated, forming a fully exfoliated nanocomposite product. However, polymer can only grow within the clay gallery when monomer effectively diffuses into the tethered radicals within the interlayer spacing. The time scale of diffusion is such that access to the monomer from within the layers is limited. Considering the rapid kinetics for free radical polymerization in solution, this intercalative monomer diffusion is much slower toward monomer addition. Thus, free initiators exhaust the monomers while SIP inside clay lamellae is delayed by diffusion. Furthermore, there is also competition from the surface-perimeter-attached initiators of the clay particles. From the combination of XRD peak position (Figure 19), low molecular weight, and higher polydispersity, we believe that most of the polymer obtained is the result of chain initiation and growth from initiators attached to the nanoparticle perimeter rather than the clay intergallery spaces. This explanation is consistent with bi-PS-M-2 retaining some intercalated structure. Even if some of the bicationic initiators were activated and grew to become oligomers, the growing chains will likely be terminated by recombination or disproportionation by nearby immobilized growing chains/initiators in the same gallery. Hence the low molecular weight and high polydispersity of bi-PS-M-2 can be explained.

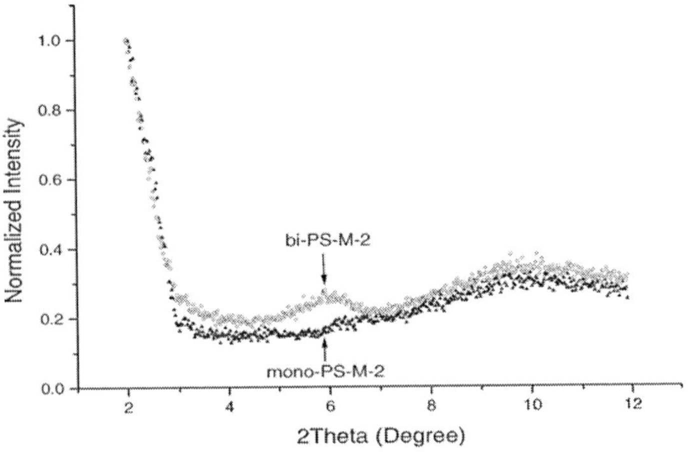


Figure 19. XRD spectra of the two SIP nanocomposites showing degree of exfoliation. [From ref. 47, reprinted with permission from American Chemical Society]

The intercalated monocationic initiator is more easily delaminated than the bicationic initiator, in comparison. The monocationic initiator molecule is also more organophilic. The weaker van der Waals interaction between the alkyl headgroups of the monocationic initiator and clay surfaces makes the intercalated clay more easily swelled by the solvent and monomer. Once the clay intercalated with monocationic initiator is exfoliated by sonicating and stirring, the attached initiators have more accessibility to monomer, that is, better monomer intercalative diffusion. On the other hand, neighboring clay platelets are electrostatically bound by the two charged ends in the case of the bicationic initiator. It is more difficult to delaminate clay stacks that are ionically held together. Thus, SIP of mono-PS-M-2 is much more successful than that of bi-PS-M-2. This explanation is supported by the higher molecular weight of its free polymer and the exfoliated structure of the product.

We had been worked to find out the reason for the appearance of XRD peak at higher 2θ value during in situ polymerization with the bicationic initiator bound clay layers. A new concept, viscosity competition between the intra- and extra-gallery phases is revealed, to explain in detail. As shown in Figure 20, the schematic diagram elucidates the possibility of less intense and broad XRD peak appeared at 2θ = 6° in higher clay loading of bicationic initiator bound clay layers. The exfoliation was only possible when the intra-gallery polymerization was at least equal to, or faster, than that of the extra-gallery polymerization

[50]. A faster intra-gallery polymerization renders the extra-gallery viscosity lower than that of the intra-gallery, facilitating diffusion of the monomer molecules into the clay galleries, allowing intra-gallery polymerization to continue. Furthermore, the lower viscosity of the extra-gallery matrix offers a lower viscous resistance against the clay layer separation from the rest of the clay tactoid.

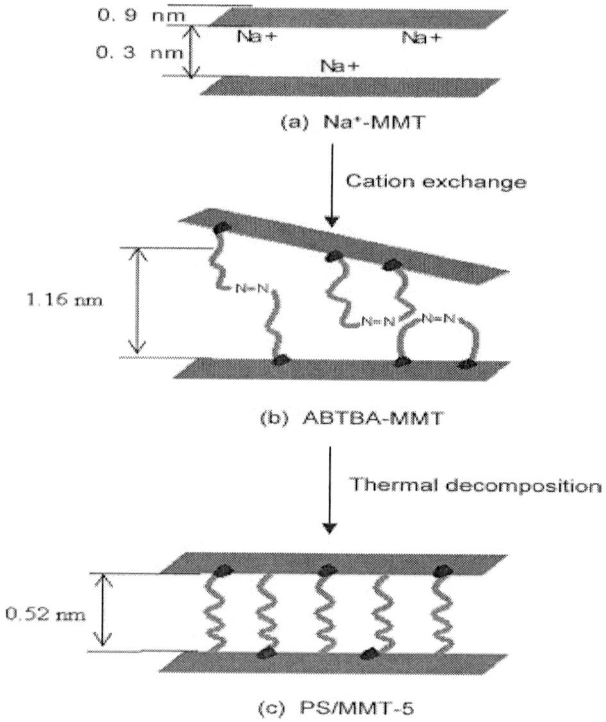


Figure 20. Schematics of the interlayer distances of (a) Na+-MMT, (b) ABTBA-MMT and (c) PS/MMT-5. [From ref. 41, reprinted with permission from Elsevier]

However, the initiator concentration in the extra-gallery phase increased with increasing clay content, as the availability of the initiator in the extra-gallery phase increased with the clay content, due to the unattached and anchored initiator molecules at the outer surfaces of the clay tactoids, which can facilitate the extra-gallery polymerization. Above a certain initiator concentration, in the extra-gallery phase, the extra-gallery polymerization becomes faster than the intra-gallery polymerization, due to the much higher monomer concentration in the resulting extra-gallery region of the clay tactoids. This stops the exfoliation of the layered silicates, as the higher viscosity of the extra-gallery matrix acts against the movement of the clay layers from the clay tactoids. Thus, it is clear that the rate of the intra-gallery polymerization is equal to, or higher, than that of the extra-gallery polymerization, when the ABTBA-MMT content was below 5 wt%. However, at 5 wt% ABTBA-MMT, the extra-gallery polymerization proceeded fast enough to provide a very high viscosity of the extra-gallery matrix, preventing the exfoliation of the silicate layers from the clay tactoids.

## 4.10. Thermal and Mechanical Properties of Polymer Clay Nanocomposites

Nanocomposites possess their unique properties due to the homogeneous dispersion of ultrafine individual silicate platelets on a nanoscale. The enhancement of the properties usually depends on the degree of dispersion and the silicate layers content within the polymer matrix. They often exhibit remarkable improvement in the tensile strength [55-58], thermal stability [59], gas barrier [52,60-62], reduced flammability [63-65] and biodegradation [66], etc., as compared with the virgin polymers, or conventional micro and macro-composites.

Table 3 provides the thermal stability of polymer-clay nanocomposites, based on the TGA measurements. The onset degradation temperatures of the PS/MMT nanocomposites increased with the clay content. The onset degradation temperature of the pristine PS was 406 °C, but that of the PS/MMT-1 was 441°C, indicating a 35°C improvement with just 1 wt% of the clay hybrid. The origin of the noticeable increase in the decomposition temperature mainly results from the unique ability of the highly dispersed nanoscale silicate layers to obstruct the volatile gas produced by the decomposition. The thermal stability was further enhanced when the clay hybrid was increased to 3 wt%, but above this limit no further improvement was observed, probably due to the increased number of unexfoliated clay tactoids. This result clearly indicated that the thermal stability of the PS/MMT nanocomposites was dependent on the clay content and exfoliation in the PS matrix [16,67]. The glass transition temperatures (Tgs) of the pure PS, PS/MMT-1, PS/MMT-3 and PS/MMT-5 were estimated to be 98.5, 103.8, 105.0 and 105.2 °C, respectively. This result indicated that the $T_g$ increased with the clay content. The monotonic improvement in the $T_g$ was due to the silicate nanoplatelets with high aspect ratios in the PS matrix. The increased $T_g$ resulted from the restricted segmental motions of the polymer chains at the organic-inorganic interface, due to the confinement of the PS chains between the silicate layers, as well as the silicate surface-polymer interaction in the nanostructured hybrids. However, no improvement in the $T_g$ was observed in the PS/MMT-5 nanocomposites. This result also indicated that the increased $T_g$ was only observed when the silicate nanoplatelets were reasonably well exfoliated and dispersed. The majority of the other well dispersed polymer nanocomposites also exhibited higher $T_g$ than their corresponding pristine polymers [68].

The effects of clay loadings on tensile properties of the PS/MMT nanocomposites are summarized in Table 4. The tensile strength and Young's modulus were significantly enhanced in the presence of the small contents of clay, while the elongation at break was reduced with increasing the clay content. The increase in tensile strength was attributed to the stronger interfacial adhesion between PS and the clay platelets. However, the enhancement of modulus was reasonably ascribed to the high resistance exerted by the clay platelets against the plastic deformation and the stretching resistance of the oriented polymer backbones in the galleries. The improvement of tensile strength in PS/MMT-3 compared to pure PS was ~ 47%, which is greater than the earlier data (~21%) for PS/MMT nanocomposite with 3 wt% MMT prepared by melt blending [69]. Similarly, the enhancement of Young's modulus in PS/MMT-3 compared to pure PS was ~25%, which is much greater than the reported value (7.4% improvement for PS/MMT nanocomposite with 5 wt% clay prepared by emulsion polymerization) [70].

**Table 3. Thermal degradation data of PS/MMT nanocomposites, based on the TGA curve**

| Sample | Degradation temperatures (°C) | | | Tg (°C) |
|---|---|---|---|---|
| | Initial | Maximum | End | |
| Pure PS | 332 | 426 | 450 | 98.5 |
| PS/MMT-1 | 361 | 458 | 497 | 103.8 |
| PS/MMT-3 | 394 | 478 | 524 | 105.0 |
| PS/MMT-5 | 390 | 483 | 509 | 105.2 |

**Table 4. Tensile properties of the PS/MMT nanocomposites**

| Sample | Tensile strength (MPa) | Young's modulus (Mpa) | Elongation at break (%) |
|---|---|---|---|
| Pure PS | 29 | 853 | 11 |
| PS/MMT-1 | 38 | 962 | 11 |
| PS/MMT-3 | 44 | 1030 | 10 |
| PS/MMT-5 | 42 | 1069 | 6 |

## 4.11. Summary

Several cationic azo-initiators containing quaternary ammonium salts are designed for the preparation of initiator-bound clay layers, by ion exchange method. Complete replacement of sodium ions in the clay galleries with the initiator cations are confirmed by the various techniques including XRD, XPS, FT-IR. The interlayer d-spacing of the resulting initiator-bound clays are largely dependent on the size and their orientation of the organic initiators. The competence between the monocationic and bicationic initiators toward the surface functionalization upon ion exchange is discussed. Monocationic initiators show higher d-spacing than that of bicationic initiators and hence the monocationic initiator is more easily delaminated when compared to bicationic initiator, during the polymerization. At the same, larger d spacing resulted in bicationic initiator-bound clay also controlled by the chain length and diameter of initiator molecules, leading to the formation of exfoliated PS/clay nanocomposites with lesser amount bicationic initiator-bound clay layers. The successful exfoliation of polymer clay nanocomposites resulted from the polymer grafting from the clay surface makes superiority over the other techniques. Hence, they offered a very high improvement in both thermal and mechanical properties, due to the high aspect ratio of individual clay layers that are uniformly dispersed throughout the polymer matrix.

# References

[1] Bailey, SW. *Reviews in Mineralogy*; Virginia Polytechnic Institute and State University: Blacksburg, 1984.
[2] Bailey, SW. *Crystal Structure of Clay Minerals and Their X-Ray Identification*; Mineralogical Society: *London*, 1980.
[3] Usuki, A; Kojima, Y; Kawasumi, M; Okada, A; Fukushima, Y; Kurauchi, T; Kamigaito, *OJ. Mater. Res*., 1993, 8, 1174.
[4] Usuki, A; Kojima, Y; Kawasumi, M; Okada, A; Kurauchi, T; Kamigaito, O. *J. Mater. Res*., 1993, 8, 1179.
[5] Okada, A; Kawasumi, M; Kurauchi, T; Kamigaito, O. *Polym. Prepr*., 1987, 28, 447.
[6] Kojima, Y; Usuki, A; Kawasumi, M; Okada, A; Kurauchi, T; Kamigaito, O. J. Poly. Sci., *Part A.: Polym. Chem*., 1993, 31, 983.
[7] Kojima, Y; Usuki, A; Kawasumi, M; Okada, A; Fukushima, Y; Kurauchi, T; Kamigaito, O. *J. Mater. Res*., 1993, 8, 1185.
[8] Lan, T; Kaviratna, PD; Pinnavaia, TJ. *J. Polym. Mater. Sci. Eng*., 1994, 71, 527.
[9] Messersmith, PB; Giannelis, EP. *Chem. Mater*., 1994, 6, 1719.
[10] Lan, T; Pinnavaia, TJ. *Chem. Mater*., 1994, 6, 2216.
[11] Fan, X; Xia, C; Advincula, RC. Colliods Surf. A: Physicochem. *Eng. Aspects*, 2003, 219, 75.
[12] Weimer, MW; Chen, H; Giannelis, EP; Sogah, DY. *J. Am. Chem. Soc*., 1999, 121, 1615.
[13] Uthirakumar, P; Kim, CJ; Nahm, KS; Hahn, YB; Lee, YS. *Colloids Surf. A: Physicochem. Eng. Aspects*, 2004, 247, 69.
[14] Tseng, CR; Wu, JJ; Lee, HY; Chang, FC. *J. Appl. Polym. Sci*., 2002, 85, 1370.
[15] Fu, X; Qutubuddin, S. *Mater. Lett*, 2000, 42, 12.
[16] Fu, X; Qutubuddin, S. *Polymer,* 2001, 42, 807.
[17] Thomas, GW. American Society of Agronomy, Inc. and Soil Science of America, Inc.: Madison, WI, 1982; pp 159.
[18] Rytwo, G; Serban, C; Nir, S; Margulies, L. *Clays Clay Miner*., 1991, 39, 551.
[19] Mittal, VJ. *Colloid Interface Sci*., 2007, 314,141.
[20] Vaia, RA; Jandt, KD; Kramer, EJ; Giannelis, EP. *Chem. Mater*., 1996, 8, 2628.
[21] Maiti, P; Nam, PH; Okamoto, M; Hasegawa, N; Usuki, A. *Macromolecules*, 2002, 35, 2042.
[22] Hasegawa, N; Okamoto, H; Kato, M; Usuki, AJ. *Appl. Polym. Sci*., 2000, 78, 1918.
[23] Ma, J; Xu, J; Ren, JH; Yu, ZZ; Mai, YW. *Polymer*, 2003, 44, 4619.
[24] Jimenez, G; Ogata, N; Kawai, H; Ogihara, TJ. *Appl. Polym. Sci*., 1997, 64, 2211.
[25] Ogata, N; Jimenez, G; Kawai, H; Ogihara, T. *J. Polym. Sci., Part B: Polym. Phys*., 1997, 35, 389.
[26] Hwu, JM; Ko, TH; Yang, WT; Lin, JC; Jiang, GJ; Xie, W; Pan, WP. *J. Appl. Polym. Sci*., 2004, 91, 101.
[27] Chen, GM; Ma, YM; Qi, ZN. *Scr. Mater*., 2001, 44, 125.
[28] Kim, YK; Choi, YS; Wang, KH; Chung, I. *J. Chem. Mater*., 2002, 14, 4990.
[29] Zhu, J; Morgan, AB; Lamelas, FJ; Wilkie, CA. *Chem. Mater*., 2001, 13, 3774.

[30] Yeh, J.-M; Liou, S.-J; Lin, C.-Y; Cheng, C.-Y; Chang,Y.-W; Lee, K.-R. *Chem. Mater.* 2002, 14, 154.

[31] Meier, L; Shelden, R; Caseri, W; Suter, U. *Macromolecules*, 1994, 27, 1637

[32] Velten, U; Shelden, RA; Caseri, WR; Suter, UW; Li, Y. *Macromolecules*, 1999, 32, 3590.

[33] Velten, U; Tossati, S; Shelden, R. A; Caseri, WR; Suter, UW; Hermann, R; Muller, M. *Langmuir*, 1999, 15, 6940.

[34] Fan, X; Xia, C; Fulghum, T; Park, MK; Locklin, J; Advincula, RC. *Langmuir*, 2003, 19, 916.

[35] Su, S; Wilkie, CA. *J. Polym. Sci., Part A: Polym. Chem.*, 2003, 41, 1124.

[36] Moet, A. S; Akelah, A. *Mater. Lett*, 1993, 18, 97.

[37] Albrecht, M; Ehrler, S; Muhlebach, A. *Macromol. Rapid Commun*, 2003, 24 , 382.

[38] Prucker, O; Ruhe, J. *Macromolecules*, 1998, 31, 602.

[39] Prucker, O; Ruhe, J. *Macromolecules*, 1998, 31, 592.

[40] Biasci, L; Aglietto, M; Ruggeri, G; Ciardelli, F. *Polymer*, 1994, 35, 3296.

[41] Uthirakumar, P; Nahm, KS; Hahn, YB; Lee, YS. *Eur. Polym. J.*, 2004, 40, 2437.

[42] Uthirakumar, P; Song, MK; Nah, C; Lee, YS. *Eur. Polym. J.*, 2005, 41, 211.

[43] Uthirakumar, P; Hahn, YB; Nahm, KS; Lee, YS. *Eur. Polym. J.*, 2005, 41, 1582.

[44] Uthirakumar, P; Kim, HJ; Hong, CH; Suh, EK; Lee, YS. Polym. *Comp.* 2008, 29, 142.

[45] Zeng, C; Lee, L. J. *Macromolecules*, 2001, 34, 4098

[46] Neises, B; Steglich W. *Angew. Chem. Int. Ed.*, 1978, 17, 522.

[47] Fan, X; Xia, C; Advincula RC. *Langmuir*, 2003, 19, 4381.

[48] Fan, X; Xia, C; Advincula R. C. *Langmuir*, 2005, 21, 2537.

[49] Kotov, N. A; Haraszti, T; Turi, L; Zavala, G; Geer, RE; Dekany, I; Fendler, JH. J. *Am. Chem. Soc.*, 1997, 119, 6821.

[50] Park, JM; Jana, SC. *Macromolecules*, 2003, 36, 2758.

[51] Choi, Y. S; Wang, K. H; Xu, M; Chung, I.J. *Macromolecules*, 2002, 14, 2936

[52] Messersmith, PB; Giannelis, EP. *J. Polym. Sci., Part A: Polym. Chem.*, 1995, 33, 1047.

[53] Tyan, H.-L; Leu, C.-M; Wei, K.-H. *Chem. Mater.*, 2001, 13, 222.

[54] Huang, X; Brittain, W. *J. Polym. Prepr.*, 2000, 41, 521.

[55] Giannelis, EP. *Adv. Mater.*, 1996, 8, 29.

[56] Giannelis, EP; Krishnamoorti, R; Manias, E. *Adv. Polym. Sci*, 1999, 138, 107.

[57] LeBaron, PC; Wang, Z; Pinnavaia, TJ. *Appl. Clay Sci.*, 1999, 15, 11.

[58] Biswas, M; Sinha, RS. *Adv. Polym. Sci.*, 2001, 155, 167.

[59] Giannelis, EP. Appl. *Organomet. Chem*, 1998, 12, 675.

[60] Xu, R; Manias, E; Snyder, AJ; Runt, J. *Macromolecules*, 2001, 34, 337.

[61] Bharadwaj, R. K. *Macromolecules*, 2001, 34, 1989.

[62] Yano, K; Usuki, A; Okada, A; Kurauchi, T; Kamigaito, O. *J. Polym. Sci., Part A: Polym. Chem*, 1993, 31, 2493.

[63] Gilman, JW. *Appl. Clay Sci.*, 1999, 15, 31.

[64] Bourbigot, S; LeBras, M; Dabrowski, F; Gilman, JW; Kashiwagi, T. *Fire Mater*, 2000, 24, 201.

[65] Gilman, JW; Jackson, CL; Morgan, AB; Harris, Jr; R; Manias, E; Giannelis, EP; Wuthenow, M; Hilton, D; Phillips, SH. *Chem. Mater*, 2000, 12, 1866.

[66] Sinha, RS; Yamada, K; Okamoto, M; Ueda, K. *Nano Lett.*, 2002, 2, 1093.
[67] Doh, JG; Cho, I. *Polym. Bull.* 1998, 41 511.
[68] Li, Y; Zhao, B; Xie, S; Xharng, S. *Polym. Int.*, 2003, 52 892.
[69] Park, CI; Choi, WM; Kim, MK; Park, OO. *J. Polym. Sci., Part B: Polym. Phys*, 2004, 42, 1685.
[70] Noh, MW; Lee, DC. *Polym. Bull.*, 1999, 42, 619.

In: Polymer Nanocomposites: Advances in Filler Surface ISBN: 978-1-60876-125-8
Editor: Vikas Mittal 

*Chapter V*

# Grafting from Clay Surfaces Using Atom Transfer Radical Polymerization

***Devon A. Shipp*** [*]
Department of Chemistry & Biomolecular Science, and Center for Advanced Materials Processing, Clarkson University, Potsdam NY, USA

## Abstract

This chapter summarizes the use of atom transfer radical polymerization (ATRP) for the preparation of polymer-layered silicate nanocomposites (PLSNs). The ATRP mechanism does not appear to be hampered by the presence of the clay, resulting in well controlled and predictable molecular weight distributions. Initiation of ATRP can be accomplished using intercalated alkyl halides, and results in polymer chains that are grafted from the silicate layers. Production of various copolymer types, such as block, gradient, graft and statistical copolymers, can be achieved. These exhibit interesting phase separation and switchable surfaces if treated with selective solvents. The resultant nanocomposites structures are, like most in situ preparation methods, often a mixture of exfoliated and intercalated structures are formed. However, given the flexibility of ATRP in making a range of polymer architectures and functionalities, it is expected that enhanced exfoliation in PLSNs can be achieved by grafting polymers from the silicate surface using ATRP.

**Keywords**: atom transfer radical polymerization, nanocomposites, layered silicates, molecular weight distribution, initiation, intercalation, copolymer, exfoliation.

## 5.1. Introduction

It has been known for some time that the physical properties of polymer materials, such as strength, flame resistance, heat stability, and abrasion resistance, can be improved by the

[*] Corresponding author: dshipp@clarkson.edu

addition of low concentrations of inorganic materials, such as silicates, metal phosphates, or complex oxides [1]. In their simplest form, such composites are made by mixing the inorganic phase into the (typically) organic polymer, or alternatively one of the two phases is synthesized in the presence of the other component that had already been formed. In either case, the major factor in whether improved properties are observed is dictated by the ability of the inorganic phase to readily mix/blend with the organic material. To improve this factor, and thus overall materials properties, significant attention over the past 2 decades has been being given to the field of nanocomposites, which appear to offer significant commercial potential [2-4].

As the name *nano*composites suggests, the size of at least one dimension of the inorganic component within the polymer matrix is (typically) ~ 1 – 100 nm. Larger sized inorganic particles have also been used in composites, however these often show a less marked improvement in the materials properties compared with nanocomposites. The former can be regarded as a mixture of two immiscible components, and while such materials may exhibit improvements in some properties (e.g. strength and modulus), there is often a reduction in others (e.g. impact resistance). [5]. As the inorganic segments become smaller they have the ability to distribute more evenly throughout the polymer and have greater surface-to-volume ratio and increase the fraction of polymer that is influenced by the presence of the inorganic surface. [3]. This can lead produceto improvements in several properties of the material, such as mechanical properties, thermal properties, diffusion barrier properties, and it can lead to materials that are generally lighter in weight since only a few percent of inorganic material is required. Thus, nanocomposites have many advantages over more conventionally filled polymer materials [2,3].

Polymer-layered silicate nanocomposites (PLSNs) represent one key nanocomposite materials that have been intensively investigated over the past 20 years or so. The primary goal in these nanocomposites is to achieve exfoliation of the silicate (clay) layers within the polymer matrix. This is expected to achieve optimal property improvement. There are several approaches that have been used in order to do this, one of which is to perform the polymerization in the presence of the layered silicate. One such polymerization is atom transfer radical polymerization (ATRP) [6-10], a method of making polymers with narrow and predictable molecular weight distributions and specific end-group functionality. It is the purpose of this chapter to describe the use of ATRP in the preparation of PLSNs.

## 5.2. Basics Structures of Layered Silicates

### 5.2.1 Layered Silicate Structure

Layered silicates (see Scheme 1) consist of two silica tetrahedral sheets fused to an edge-shared octahedral sheet (typically aluminum or magnesium hydroxide) [11]. A lattice is formed from periodic stacking of the layers, with each layer of ~ 1 nm thickness and the stack held together by weak van der Waals forces, allowing for small molecules or macromolecules to be inserted (intercalated) between the layers. The lateral dimensions vary from 300Å to several microns depending on the specific silicate, and each layer can nominally be treated as a disc. A variety of distinct silicates are either found naturally or can be synthesized and distinguished by the location and type of cations present in the oxygen framework. Common silicates include montmorillonite, hectorite, magadiite and saponite. Depending on the degree of isomorphic substitution of the tetrahedral silicon or the octahedral aluminum and magnesium, the layers possess a net charge deficiency that is typically compensated by cations, such as $Na^+$, $K^+$, $Ca^{2+}$, in the interlayer galleries. The cation exchange capacity (CEC), which is expressed in meq/100 g, is used to quantitatively describe the negative surface charge.

The interlayer spacing between the silicate layers is normally quite hydrophilic and hydrated. The interlayer spacing can be modified through the use of alkyl-ammonium or alkyl-phosphonium cations in cation exchange reactions [11]. The organic modifiers may contain a variety of functional groups, thereby providing access to a wide range of chemistries and properties. Such functionalities can include polymerizable groups (e.g. vinyl) or initiating groups, such as for ATRP, the subject polymerization of this chapter.

## 5.2.2 Structures within PLSNs

There are two idealized structures in PLSN materials (excluding immiscible systems which normally should not be regarded as nanocomposites), as depicted in Scheme 2. These are (i) intercalated materials, where the organic polymer is present within the inorganic silicate layers (i.e. a well-ordered and structured arrangement exists), and (ii) exfoliated materials, where the organic polymer host matrix contains delaminated silicate crystallites (i.e. disordered and well dispersed). Notice that in the exfoliated case the surface area between organic and inorganic is increased compared with the intercalated materials. This is speculated to lead to greater enhancement of physical properties of exfoliated composites when compared with similar intercalated materials. It should be noted, however, that many polymer-nanocomposites materials are not simply only exfoliated or just intercalated – they often tend to have mixtures of both structures.

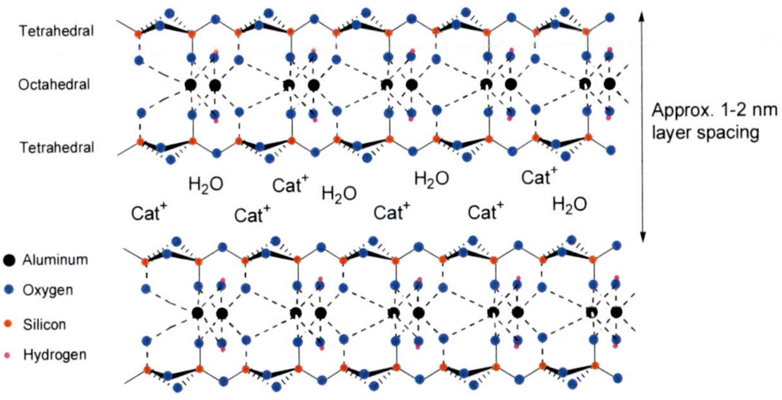


Scheme 1. Basic structure of montmorillonite, a common clay used in PLSNs. $Cat^+$ are exchangeable cations such as $Na^+$, $Ca^{2+}$, etc.

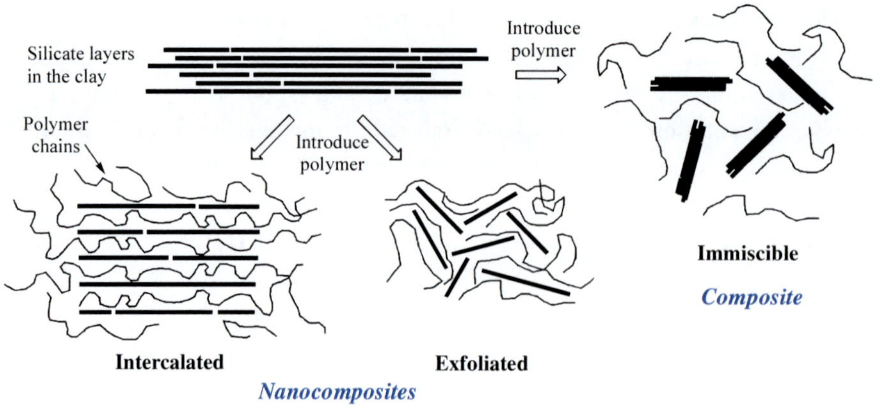


Scheme 2. Schematic of idealized accessible structures for polymer-layered silicate nanocomposites.

## 5.3 Synthetic Methods for the Preparation of PLSNs

Four general methods of synthesis for both types of nanocomposites exist. These are in situ polymerization, solution intercalation, melt intercalation (or melt blending), and sol gel techniques [12,13]. In each of these cases, the relative amounts of intercalation versus exfoliation depend on experimental conditions (e.g. monomer type, solvents, temperatures, etc.), and mixed structures are often present. While melt blending often has the advantage of being relatively straightforward in application, in situ polymerization may, in some cases, provide a wider variety of adjustable parameters, and thus access to novel materials with greater control over the final structure and property.

Approximately 15 years ago, the Toyota Central Research & Development Labs first demonstrated the improvement of several properties on the inclusion of small percentages of silicate into the polymer through in situ polymerization [14-19]. They polymerized intercalated ε-caprolactam in montmorillonite to produce a nylon 6–clay nanocomposite [14-19]. This resulted in the silicate layers being uniformly dispersed within the polymer matrix, and the composite material having improved physical properties compared with neat polymer samples (e.g. tensile strength increased by 55% with only 4.2 wt% of montmorillonite) [14]. The significance of these results is further compounded when it is noted that these improvements were obtained with only a few percent of silicate present. Other researchers have also investigated polymer–clay nanocomposites formed by in situ polymerization. Polymers have included epoxy, acrylics, polyesters, polycarbonates, vinyl polymers, polyolefins and polystyrene. Excellent reviews published by Giannelis et al [20]. Alexandre and Dubois [13], Ray and Okamoto [21] and Paul and Robeson [22], along with other chapters in this book, further cover these and other nanocomposite materials.

## 5.4 Living Radical Polymerizations & Atom Transfer Radical Polymerization

Living radical polymerizations (LRPs) give polymer materials that have well-defined molecular weights and often specific functionalities [23]. There are three primary LRP methods, including atom transfer radical polymerization (ATRP) [6-8,10], nitroxide-mediated polymerization (NMP) [24,25], and reversible addition-fragmentation chain transfer (RAFT) polymerization [26-28]. These techniques allow polymers to be produced that have well-definedpredictable molecular weights, narrow distributions (polydispersities less than 1.5, often less than 1.2) and functional groups that allow post-polymerization reactions to take place. This last attribute provides a means for polymers with novel and/or complex architecture to be synthesized. Block copolymers, star polymers, polymers attached to surfaces, hyperbranched polymers, and gradient copolymers are just a few of the new polymer materials/architectures that have been produced using ATRP, RAFT and NMP [29]. Of particular note is that these methods have been applied to the growth of polymer chains from surfaces other than clays, for example silicon, silica or gold.

ATRP (Scheme 3) has been shown to be able to polymerize a variety of monomers in a controlled manner, such as styrenics, acrylates, methacrylates, acrylonitrile and acrylamides, but has failed to control less reactive monomers that can be polymerized by a radical mechanism (e.g. vinyl acetate, olefins) [8]. The main advantage of ATRP over NMP and RAFT is the use of a relatively inexpensive chain end group or atom (typically a halogen). The major disadvantage is the use of a transition metal halide catalyst (often Cu(I)Cl, Cu(I)Br or Ru(II)$Cl_2$ complexed with a ligand) that contaminates the final polymer product.

RAFT polymerization is even more universal than ATRP in terms of monomers that can yield polymers with controlled molecular weights [26-28]. Any monomers that can be polymerized through a radical mechanism can be controlled through the RAFT process. However, the final polymer product is sometimes colored (depending on the RAFT agent used) and the odor of the thio-based RAFT agents can be a drawback. Furthermore, grafting from surfaces is also more complicated in RAFT compared with ATRP because the mechanism involves chain transfer of the radical species, which leads to more untethered chains.

NMP was the earliest LRP method developed (commonly used today), initially by the CSIRO laboratories in Australia [28,30,31], and subsequently refined by Georges et al [25], who showed that polystyrene could be produced with very narrow molecular weight distributions (polydispersities of less than 1.2) when the polymerization was conducted in the presence of the stable free radical TEMPO. However, from early work it became apparent that styrene (and a few derivatives) was the only monomer that could be effectively controlled by the most common nitroxide, TEMPO, and thus a large effort has been directed towards the discovering of more 'universal' nitroxides that would allow a wider range of monomers to be polymerized in a controlled manner [24]. These attempts have been moderately successful with acrylates, dienes and some other monomer families now being able to be polymerized in a living fashion with NMP [32].

## 5.5 Use of ATRP to Graft Polymer Chains from Clay Surfaces

The use of ATRP to graft polymer from clay (montmorillonite) surfaces was first reported by Böttcher et al [33]. They made a quaternary ammonium-based ATRP initiator (Scheme 4) which was intercalated into the sodium montmorillonite (Na-MMT) layers, then used to initiate methyl methacrylate (MMA) (Scheme 5). The resulting poly(methyl methacrylate) (PMMA) increased in number average molecular weight ($M_n$) and had quite a low polydispersity (the ratio of the weight average molecular weight, $M_w$, and the $M_n$, i.e. $M_w/M_n$), reported to be 1.06 – 1.14. There was no indication from the kinetics of the reaction that the Cu-catalyzed ATRP process was affected by the clay, as has been observed in conventional radical polymerization when clays are present [34]. X-ray diffraction (XRD) showed that the silicate layers became more dispersed as the polymerization proceeded (Figure 1), and it was claimed that complete delamination was achieved. However, no transmission electron microscopy (TEM) was used to confirm this assertion.

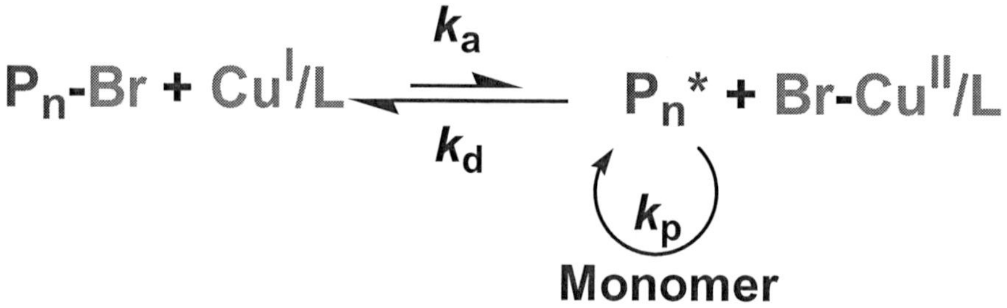


Scheme 3. Mechanism of Cu-mediated ATRP. $P_n$-Br = polymer chain bearing a Br chain end, L = ligand (often N-based), $k_a$ = activation rate constant, $k_d$ = deactivation rate constant.

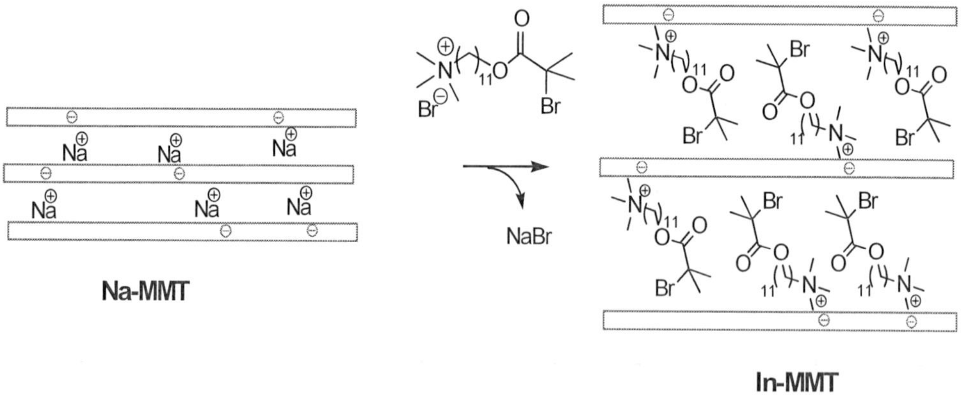


Scheme 4. Introduction of an ATRP initiator into sodium montmorillonite (Na-MMT) [33].

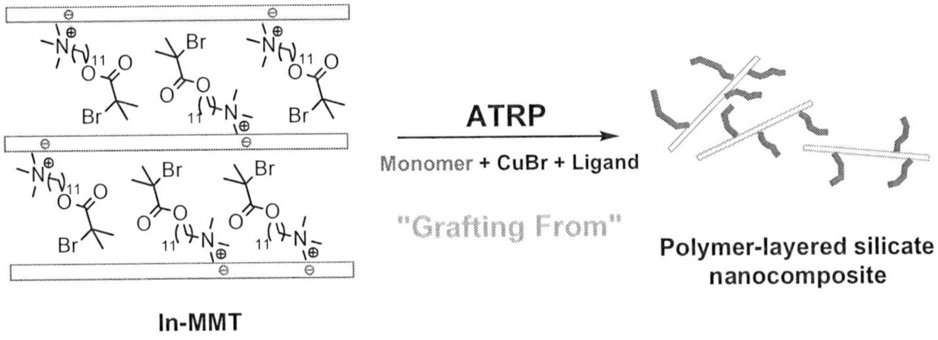


Scheme 5. Polymerization from initiator-modified montmorillonite (In-MMT) using ATRP.

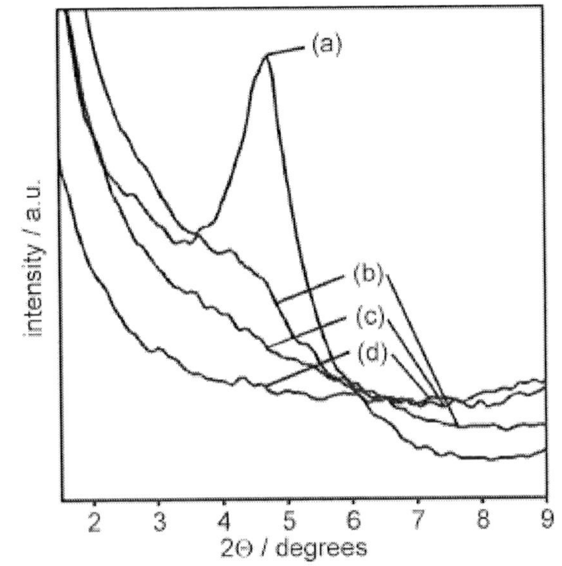


Figure 1. X-ray diffraction (in arbitrary units) versus scattering angle (in degrees): (a) initiator-modified silicate 5; (b) composite after 0.5 h time of polymerization; (c) composite after 1 h time of polymerization; (d) composite after 4 h time of polymerization. Reprinted from ref. [33], Copyright (2002), with permission from the Royal Society of Chemistry.

Shipp et al. [35] followed subsequently usedby following a similar procedure for the intercalation of an ATRP initiator into the interlayer spacing of montmorillonite, but broadened the range of monomers used and also made several block copolymer-clay nanocomposites. As shown in Figure 2, the ATRP of styrene from the intercalation ATRP initiator showed all the characteristics of a regular LRP, viz. linear increase in $M_n$ as a function of conversion, low polydispersity, and functional end groups. This last piece of information could be determined from the [[1]] H NMR spectrum of the polystyrene extracted from the nanocomposite (Figure 3).

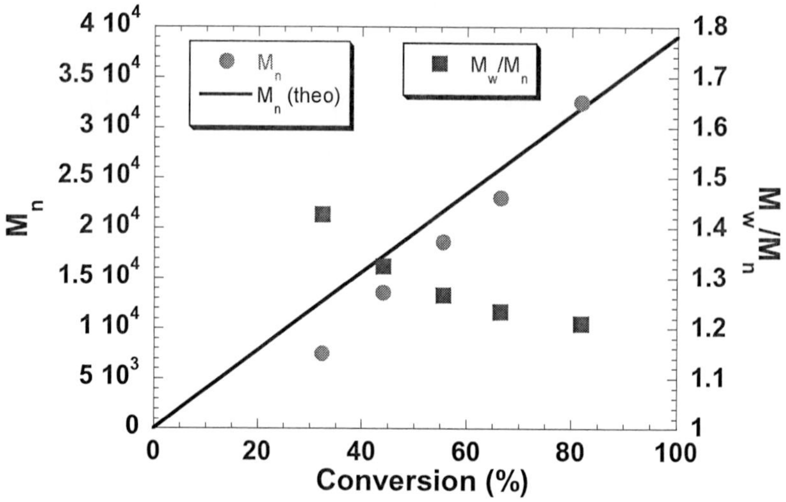


Figure 2. $M_n$ and $M_w/M_n$ as a function of styrene conversion for the ATRP of styrene from initiator modified montmorillonite [35].

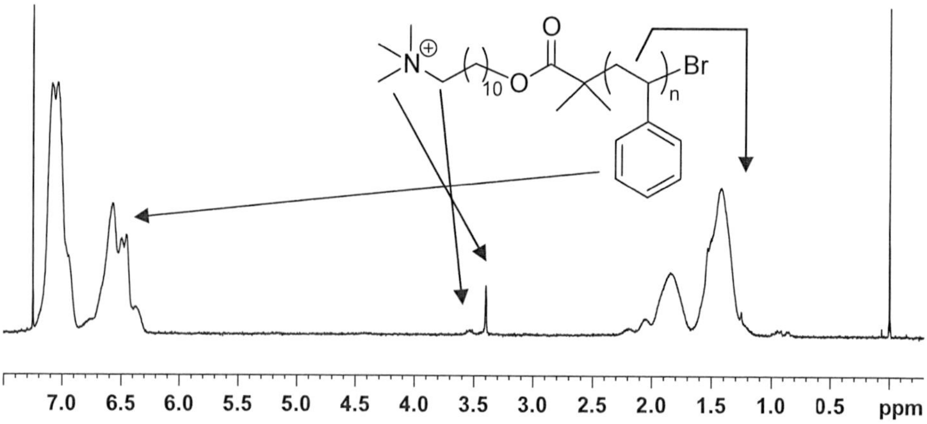


Figure 3. $^1$H NMR spectrum of the polystyrene extracted from the PS-silicate nanocomposite [35].

In the spectrum shown in Figure 3, not only isare the regular peaks from the polystyrene seen (aromatic at $\delta \sim 6.2 - 7.2$ ppm; backbone CH-$CH_2$ at $\delta \sim 1.0 - 2.2$ ppm), but also the methyl and methylene groups adjacent to the N atom in the initiator (at 3.4 ppm and 3.5 ppm, respectively). XRD of samples taken at different times during the polymerization of styrene showed the peak at $2\theta \sim 3.4^o$, which is due to a regular layered silicate structure, becoming smaller, indicating the loss of structure in the silicate (Figure 4). However, TEM showed that complete exfoliation was not achieved in the PS samples. On the other hand, exfoliation appeared to be complete in the PMMA-based samples (Figure 5).

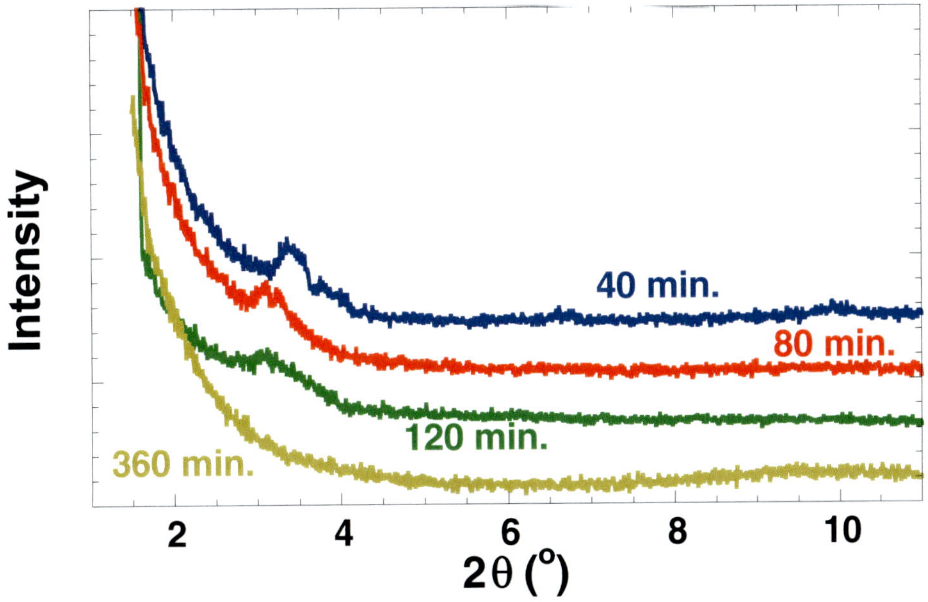


Figure 4. XRD patterns of PS-silicate nanocomposite at different ATRP reaction times [35].

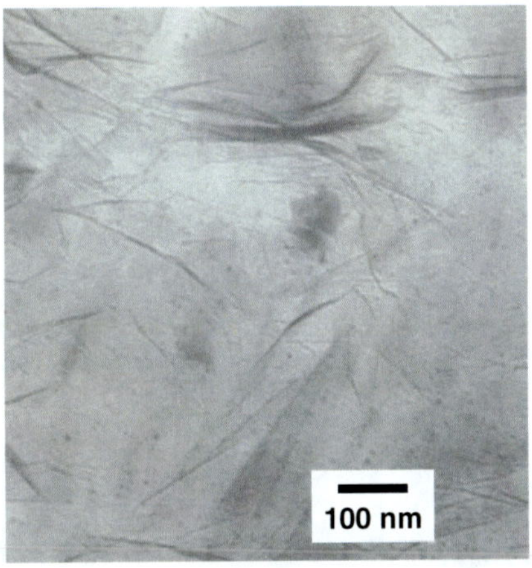


Figure 5. TEM image of a PMMA-silicate nanocomposite made by in situ ATRP [35].

Subsequently, other researchers have utilized ATRP to make homopolymer-silicate nanocomposites. Li et al. [36] made PS-magadiite nanocomposites by covalently attaching an amino silane to the hydroxyl groups on the magadiite surface, and then reacting an ATRP initiating group (2-bromopropionate) to the amine. The molecular weight distribution of the PS was controlled (polydispersitics of 1.35 or less), and the nanocomposite featured a mix of intercalated and partially exfoliated silicate layers. Mechanical properties were shown to increase as the weight percentage of magadiite increased.

Wheeler et al. [37] grafted PMMA from laponite using an ATRP initiator covalently attached to the silanol groups on the edge of the clay layers (Scheme 6). They were also able to add ammonium flat surfactant to the flat surfaces of the laponite through ion exchange, and thus compared these two grafting methods, in addition to combining them. It was found that, while in all cases there was poor dispersion of the laponite in the polymer matrix, grafting PMMA from the edges provided more stable dispersions of the clay (Figure 6) and improved the modulus by 50% at 5 wt%, compared with 25% improvement with 3 wt% of the ion exchanged nanocomposite.

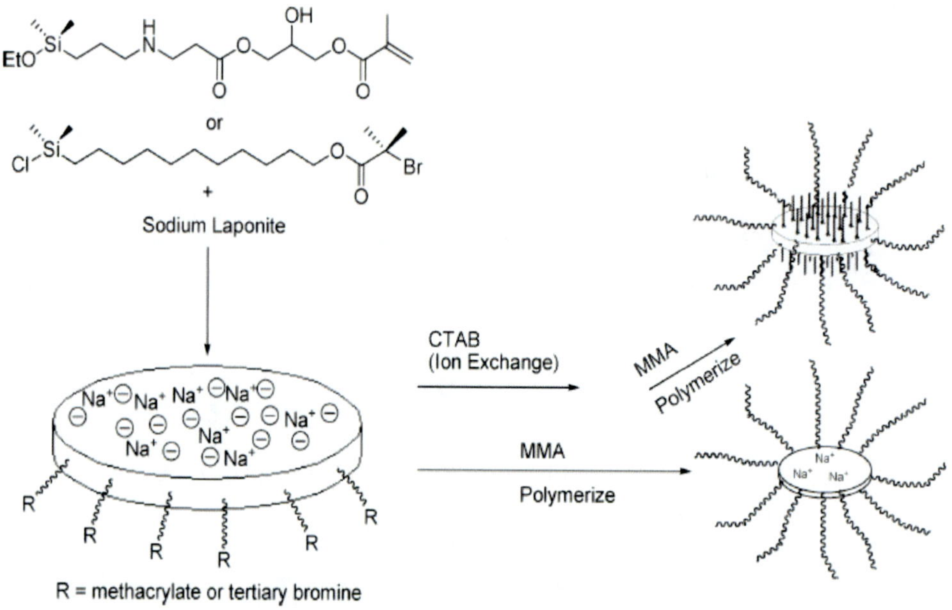


Scheme 6. Process in which polymer was covalently attached to laponite. Reprinted from ref. [37], Copyright (2006), with permission from the American Chemical Society.

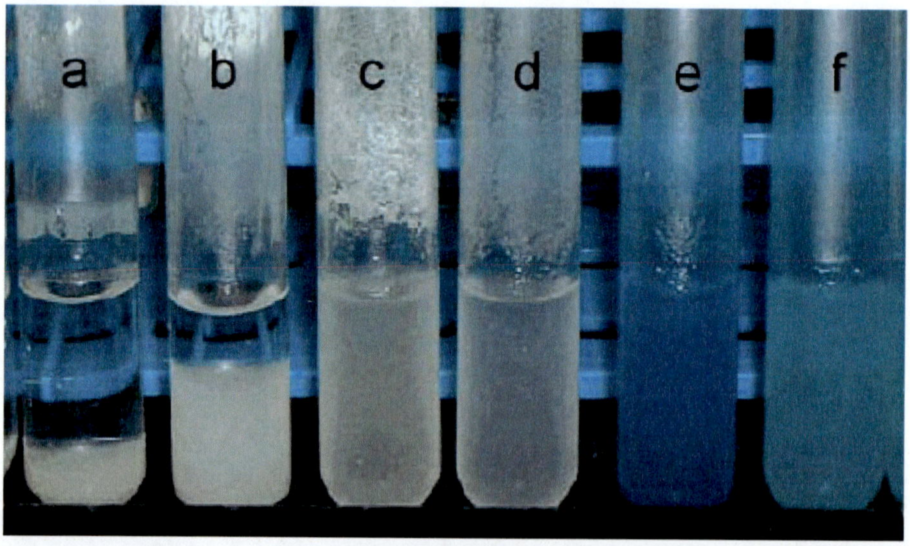


Figure 6. Clays dispersed in chloroform after allowing to settle for 2 h. (a) sodium Laponite, (b) CTAB, (c) MA/PMMA, (d) MA/CTAB/PMMA, (e) ATRP/PMMA, and (f) ATRP/CTAB/PMMA Reprinted from ref. [37], Copyright (2006), with permission from the American Chemical Society.

Qiu et al. [38] anchored an ATRP initiator inside the interlayer galleries of Zn, Al-layered double hydroxide (LDH) followed by polymerization of styrene. An increase in the thermal stability of the polystyrene-LDH nanocomposite was observed. Very recently, Oral et al. [39] polymerized 2-hydroxyethyl methacrylate (HEMA) in the presence of an ATRP-initiator functionalized montmorillonite clay.

## 5.6. Block Copolymer–Clay Nanocomposites using ATRP

Where the use of ATRP, or any of the LRP methods, is far more advantageous than conventional radical polymerization is in the production of block copolymers. Zhao and Shipp [40] were the first to make block copolymer-PLSN materials, focusing on polystyrene-block-poly(n-butyl acrylate) (PS-PBA) materials. This was made in a two-step polymerization (Scheme 7); first the PS (M1) segment was made, then the unreacted styrene was removed by vacuum and the n-butyl acrylate (M2) was added, yielding the second segment. The nanocomposite was a mixture of exfoliated and intercalated silicates, but TEM of the exfoliated regions (Figure 7) showed that the PS and PBA segments underwent the phase separation expected of these systems. It was also shown that this block copolymer system could behave as a switchable brush if selective solvents were applied [41]. Similar behavior of amphiphilic poly(methyl methacrylate-block-2-(dimethylamino)ethyl methacrylate) (PMMA-b-PDMAEMA) block copolymer brushes, which synthesized by in situ ATRP on the surface of clay layers, was observed by Zhao et al [42].

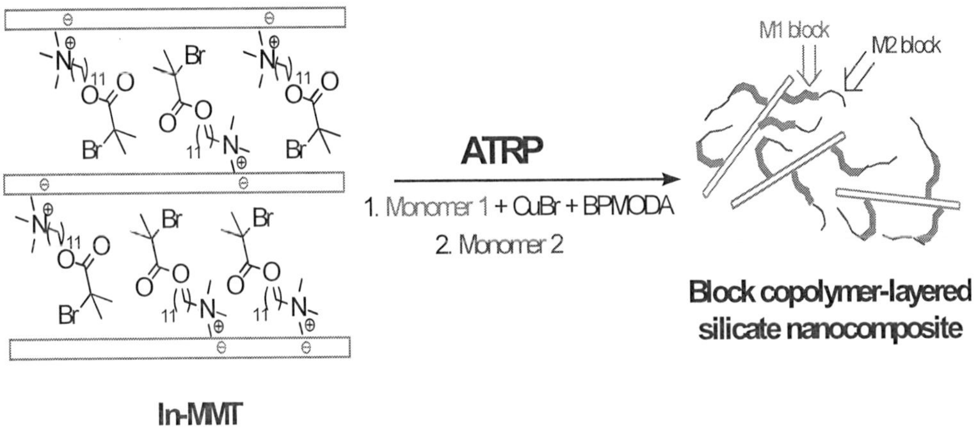


Scheme 7. 2-step approach to the synthesis of block copolymer PLSNs.

Liu et al. [43] synthesized polystyrene-block-polyacrylamide block copolymers from the surface of a fibrillar clay, attapulgite. ATRP was used under ultrasonic conditions. Zhao et al. [44] made an interesting comb-brush copolymer of poly(L-lactide) (PLLA) (the comb) on poly(2-hydroxyethyl methacrylate) (PHEMA) backbone (Scheme 8). The PHEMA was first polymerized using an ATRP initiator anchored between the montmorillonite layers, and then the hydroxyl group used to initiate ring opening polymerization of L-lactide, resulting in the comb-brush copolymer nanocomposite. A mixture of exfoliated and intercalated silicates were observed, with an increase in exfoliation being observed when the PLLA segment molecular weight was increased. Using TEM they were able to observe the copolymer on the silicate surface, and after removing the clay and depositing the brush copolymer onto a mica surface, atom force microscopy could be used to visualize the single copolymer molecules (Figure 8).

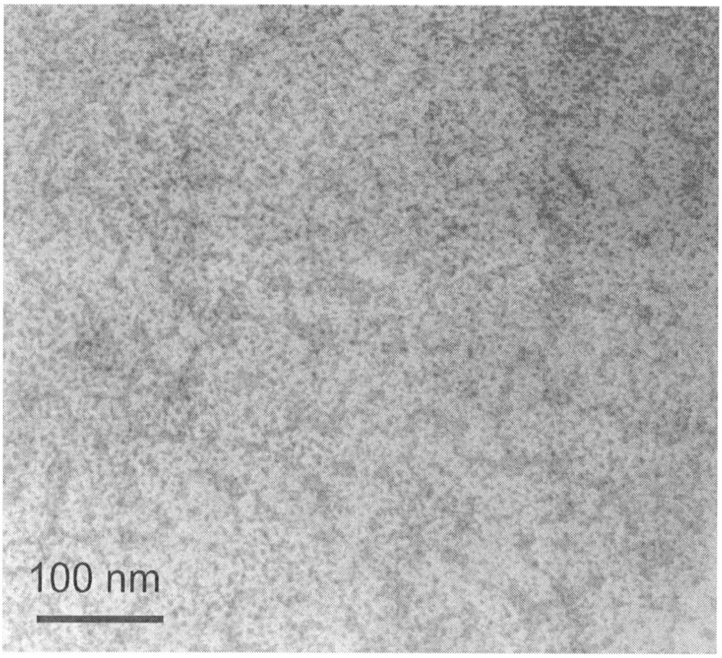


Figure 7. TEM of the PS-PBA nanocomposite in an exfoliated region, after staining (dark areas = PBA, light areas = PS). Reprinted from ref. [40], Copyright (2003), with permission from the American Chemical Society.

Liu et al. [43] synthesized polystyrene-block-polyacrylamide block copolymers from the surface of a fibrillar clay, attapulgite. ATRP was used under ultrasonic conditions. Zhao et al. [44] made an interesting comb-brush copolymer of poly(L-lactide) (PLLA) (the comb) on poly(2-hydroxyethyl methacrylate) (PHEMA) backbone (Scheme 8). The PHEMA was first polymerized using an ATRP initiator anchored between the montmorillonite layers, and then the hydroxyl group used to initiate ring opening polymerization of L-lactide, resulting in the comb-brush copolymer nanocomposite. A mixture of exfoliated and intercalated silicates were observed, with an increase in exfoliation being observed when the PLLA segment molecular weight was increased. Using TEM they were able to observe the copolymer on the silicate surface, and after removing the clay and depositing the brush copolymer onto a mica surface, atom force microscopy could be used to visualize the single copolymer molecules (Figure 8).

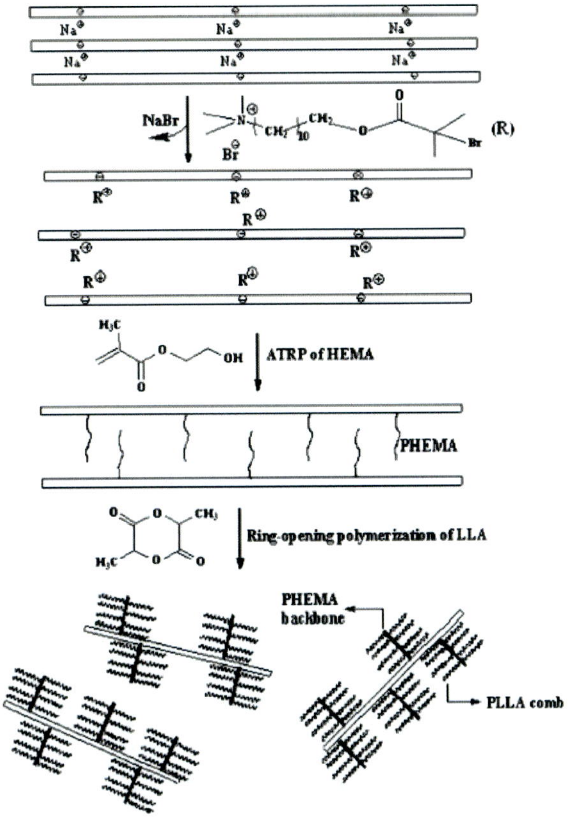


Scheme 8. Approach to making PHEMA-comb-PLLA copolymers from layered silicates. Reprinted from ref. [44], Copyright (2006), with permission from Elsevier.

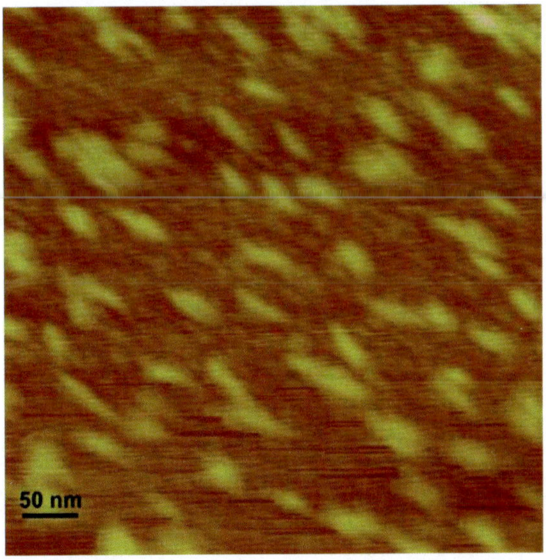


Figure 8. A tapping mode AFM image of cleaved $PLLA_{120}$-PHEMA comb polymer. The sample was prepared on a freshly cleaved mica. Reprinted from ref. [44], Copyright (2006), with permission from Elsevier.

## Conclusions

To summarize, ATRP has been shown to be very useful in the preparation of PLSNs. The ATRP mechanism does not appear to be hampered by the presence of the clay, resulting in well controlled and predictable molecular weight distributions. While the nanocomposites do not necessarily become exfoliated using this approach (like most preparation methods, it more often than not results in a mixture of exfoliated and intercalated structures), where ATRP has been exceedingly useful is in the preparation of novel block copolymers. These exhibit interesting phase separation and switchable surfaces if treated with selective solvents.

While there are many more materials that could be made using a combination of ATRP and layered silicates, there is still more work needed to be done in order to examine what the effect of functionality and polymer structure can have on silicate layer dispersion. As noted above, most examples cited in this chapter give a mixture of exfoliated and intercalated structures – perhaps with the right architecture and composition a PLSN master-batch could be made by ATRP and subsequently blended into other polymers to give significantly exfoliated nanocomposites. In this regard, it is interesting to note that theoretical work has predicted that star polymers might provide a better route to exfoliated PLSNs. [45]. Robello et al. [46] followed this prediction up with experimental evidence that indeed this is the case (they used ATRP to make the 5-arm star first, and then added the star polymer to the layered silicate). It may be that utilizing the flexibility of ATRP in making a range of polymer architectures that enhance exfoliation in PLSN – more work in this area is warranted in order to examine such structure-property relationships in more detail.

## Acknowledgements

It is with pleasure that I acknowledge those that I have been fortunate to work with in the area of ATRP and PLSN, including S.D. Argoti, B.P. Farrell, and in particular I greatly appreciate the excellent collaborations with Prof. Hanying Zhao. I am also indebted to various funding agencies and corporations for their support of our research, including Kuraray America Inc., the donors of the Petroleum Research Fund (administered by the American Chemical Society), and the Center for Advanced Materials Processing (CAMP), a New York State Center for Advanced Technology.

I dedicate this work to Professor David H. Solomon, whose work in so many areas has inspired many generations of chemists and materials scientists around the world, and particularly, not only in Australia, but around the world. He hias influenced is felt in so many technological areas, not the least in the two areas discussed in this chapter: the chemistry of clays (see, for example, reference 34) and radical polymerization, particularly living radical polymerization (reference 31 and as cited in reference 28).

# References


[1] Mark, JE. *Polym. Eng. Sci*, 1996, 36, 2905-2920.

[2] Balazs, AC; Emrick, T; Russell, TP. *Science*, 2006, 314, 1107-1110.

[3] Crosby, AJ; Lee, J.-Y. *Polym. Rev.*, 2007, 47, 217-229.

[4] In *Polymer-Clay Nanocomposites*; Pinnavaia, TJ; Beall, GW; Eds; John Wiley & Sons: New York, 2000.

[5] Giannelis, EP. *Adv. Mater.*, 1996, 8, 29-35.

[6] Kamigaito, M; Ando, T; Sawamoto, M. *Chem. Rev.* 2001, 101, 3689-3746.

[7] Kato, M; Kamigaito, M; Sawamoto, M; Higashimura, T. *Macromolecules*, 1995, 28, 1721-1723.

[8] Matyjaszewski, K; Xia, J. *Chem. Rev.*, 2001, 101, 2921-2990.

[9] Tsarevsky, NV; Matyjaszewski, K. *Chem. Rev.*, 2007, 107, 2270-2299.

[10] Wang, JS; Matyjaszewski, K. *J. Am. Chem. Soc.*, 1995, 117, 5614-5615.

[11] Pinnavaia, T. J. *Science*, 1983, 220, 365-371.

[12] Zanetti, M; Lomakin, S; Camino, G. *Macromol. Mater. Eng.*, 2000, 279, 1-9.

[13] Alexandre, M; Dubois, P. *Mater. Sci. Eng.*, 2000, R28, 1-63.

[14] Kojima, Y; Usuki, A; Kawasumi, M; Okada, A; Fukushima, Y; Kurauchi, T; Kamigaito, O. *J. Mater. Res.*, 1993, 8, 1185-1189.

[15] Kojima, Y; Usuki, A; Kawasumi, M; Okada, A; Karauchi, T; Kamigaito, O. *J. Polym. Sci., Part A: Polym. Chem.*, 1993, 31, 983-986.

[16] Kojima, Y; Usuki, A; Kawasumi, M; Okada, A; Kurauchi, T; Kamigaito, O; Kaji, K. *J. Polym. Sci., Part B: Polym. Phys.*, 1995, 33, 1039-1045.

[17] Usuki, A; Kato, M; Okada, A; Kurauchi, T. *J. Appl. Polym. Sci.*, 1997, 63, 137-139.

[18] Usuki, A; Kawasumi, M; Kojima, Y; Okada, A; Kurauchi, T; Kamigaito, O. *J. Mater. Res*, 1993, 8, 1174-1178.

[19] Usuki, A; Kojima, Y; Kawasumi, M; Okada, A; Fukushima, Y; Kurauchi, T; Kamigaito, O. *J. Mater. Res*, 1993, 8, 1179-1184.

[20] Giannelis, EP; Krishnamoorti, R; Manias, E. *Adv. Polym. Sci*, 1999, 138, 107-147.

[21] Ray, SS; Okamoto, M. *Prog. Polym. Sci*, 2003, 28, 1539-1641.

[22] Paul, DR; Robeson, LM. *Polymer*, 2008, 49, 3187-3204.

[23] Shipp, DA. *J. Macromol. Sci., Part C: Polym. Rev.*, 2005, 45, 171-194.

[24] Hawker, CJ; Bosman, AW; Harth, E. *Chem. Rev*, 2001, 101, 3661-3688.

[25] Georges, MK; Veregin, RPN; Kazmaier, PM; Hamer, GK. *Macromolecules*, 1993, 26, 2987-2988.

[26] Moad, G; Rizzardo, E; Thang, S. *Aust. J. Chem*, 2005, 58, 379-410.

[27] In *Handbook of RAFT Polymerization*; Barner-Kowollik, C; Ed; Wiley-VCH: Weinheim, Germany, 2008.

[28] Moad, G; Rizzardo, E; Thang, SH. *Acc. Chem. Res*, 2008, 41, 1133-1142.

[29] In *Handbook of Radical Polymerization*; Matyjaszewski, K; Davis, T. P; Eds; Wiley & Sons: Hoboken, NJ, 2002.

[30] Rizzardo, E. *Chem. Aust*, 1987, 54, 32.

[31] Solomon, DH; Rizzardo, E; Cacioli, P. *U.S. Pat.*, 4,581,429 1986.

[32] Sciannamea, V; Jérôme, R; Detrembleur, C. *Chem. Rev*, 2008, 108, 1104–1126.

[33] Böttcher, H; Hallensleben, ML; Nuss, S; Wurm, H; Bauer, J; Behrens, P. *J. Mater. Chem*, 2002, 12, 1351-1354.

[34] Solomon, DH; Swift, JD. *J. Appl. Polym. Sci*, 1967, 11, 2567-2575.

[35] Zhao, HY; Argoti, SD; Farrell, BP; Shipp, DA. *J. Polym. Sci., Part A: Polym. Chem*, 2004, 42, 916-924.

[36] Li, CP; Huang, CM; Hsieh, MT; Wei, KH. *J. Polym. Sci., Part A: Polym. Chem*, 2005, 43, 534-542.

[37] Wheeler, PA; Wang, JZ; Mathias, L. J. *Chem. Mater.*, 2006, 18, 3937-3945.

[38] Qiu, LZ; Chen, W; Qu, B. *J. Coll. Polym. Sci.*, 2005, 283, 1241-1245.

[39] Oral, A; Shahwan, T; Guler, C. *J. Mater. Res.*, 2008, 23, 3316-3322.

[40] Zhao, H; Shipp, DA. *Chem. Mater.* 2003, 15, 2693-2695.

[41] Zhao, HY; Farrell, BP; Shipp, DA. *Polymer*, 2004, 45, 4473-4481.

[42] Yang, YF; Liu, L; Zhang, J; Li, C. X; Zhao, H. Y. *Langmuir* 2007, 23, 2867-2873.

[43] Liu, P; Wang, TM; Su, ZX. *J. Nanosci. Nanotechnol.*, 2006, 6, 1684-1688.

[44] Yang, YF; Wu, DX; Li, CX; Liu, L; Cheng, XH; Zhao, H. Y. *Polymer*, 2006, 47, 7374-7381.

[45] Singh, C; Balazs, AC. *Polym. Int.* 2000, 49, 469-471.

[46] Robello, DR; Yamaguchi, N; Blanton, T; Barnes, C. *J. Am. Chem. Soc.*, 2004, 126, 8118-8119.

In: Polymer Nanocomposites: Advances in Filler Surface
Editor: Vikas Mittal
ISBN: 978-1-60876-125-8


*Chapter VI*

# Exchange of Functional Modifications on the Clay Surface

***Jannick Duchet-Rumeau*** *

Université de Lyon, F-69003, Lyon, France INSA Lyon, F-69621, Villeurbanne, France CNRS, UMR5223, Ingénierie des Matériaux Polymères

## Abstract

In this chapter, we have discussed the cationic exchange process which is the ordinary procedure to adsorb a chemical containing a charge-bearing group (i.e. an ionized molecule) on the internal surface of clay galleries. If the adsorbate is an organic molecule containing a charge-bearing group with a high affinity for clay immobilized anions (e.g. a quaternary ammonium salt), the cationic exchange process modifies clay nature (from hydrophilic to hydrophobic) to an extent which depends on the nature of the organic molecule. This treatment generates organically-modified clays, characterized by an improved compatibility with most of the polymers. If the adsorbate bears more than one function (e.g. cationic fluorescent dyes contain charge-bearing any fluorophore-bearing groups), cationic exchange process is a suitable method to render clays photo-active. Photofunctionalization can open large possibilities for all the modern on-line process monitoring techniques by conceiving a detection system capable of perform real-time monitoring of melt compounding of a polymer and a lamellar filler and sensitive to multiscale morphologies.

**Keywords**: ionized molecule, quaternary ammonium salt, adsorbate, photofunctionalization, nanocomposites, morphology, hydrophobic and hydrophilic clays.

---

* Corresponding author: Jannick.duchet@insa-lyon.fr

## 6.1. Introduction

The addition of clays (hydrophilic by nature) to polymers (mostly hydrophobic) may greatly improve their properties, but in case the particles are perfectly disaggregated, dispersed and distributed into the matrix. Cationic exchange processes are commonly used to improve clay compatibility with polymers, thus facilitating a more intimate mixture of the fillers with the matrix. Many clays are aluminosilicates with a layered structure that contain typically two types of sheets, tetrahedral (T) and octahedral (O). The T sheets are made of silicon atoms tetrahedrally coordinated by four oxygen atoms and the O sheets have metallic cations (usually aluminium or magnesium) coordinated with six oxygens or hydroxyls. T and O sheets are covalently linked through shared oxygens in a ratio 1:1 (TO) or 2:1 (TOT). A 2:1 ratio results in smectite clays, the most common of which is montmorillonite (see Figure 1). The wide variety of clays found in nature occurs due to the myriad substitutions that can occur in T or O sites. Al(III) can replace Si(IV) in the tetrahedral sheet or Mg(II) for Al(III) in the octahedral layers and therefore cause a negative charge. This permanent negative charge due to isomorphous substitutions is equilibrated by the presence of singly- and doubly-charged mobile cations ($Na^+$, $K^+$ and $Ca^{2+}$, $Mg^{2+}$) which act as *counterions* (*i.e.* compensating cations). These counterions are principally grouped in the interlayer space (also known as *clay gallery*) – which separates the platelets and whose dimension (*basal spacing* $d_{001}$) depends on the nature of the cations and the degree of clay hydration –but some of them are also located on the external surface of the platelets. The *surface counterions* can be readily exchanged with other inorganic or organic ions in all clay minerals, whilst the *interlayer counterions* are only accessible in *expandable* (*swelling*) *clays*. One important consequence of the charged nature of the clays is that they are generally highly hydrophilic species and therefore naturally incompatible with a wide range of polymer types. A necessary prerequisite for successful formation of polymer-clay nanocomposites is therefore alteration of the clay polarity to make the clay "organophilic". This can be readily achieved through ion-exchange reactions which is the most used modification method to modify clay nature (from hydrophilic to hydrophobic) to an extent which depends on the nature of the organic molecule. This surface treatment generates organically-modified clays (or organoclays) characterized by an improved compatibility with most of the polymers. If the adsorbate bears more than one function, cationic exchange process is still possible and suitable to functionalize clays, for example by tailoring their optical properties. The aim of this chapter is to review and summarize the main surface treatments by cationic exchange to bring to the lamellar silicates a specific function.

## 6.2. The Process of Cationic Exchange (CE)

Gieseking [GIE39] and Hendricks [HEN41] showed fort the first time that inorganic exchangeable cations of clay minerals can be replaced by organic cations through ion exchange reactions in aqueous solution of organic-base salts, such as hydrochlorides. The reaction that takes place is the following:

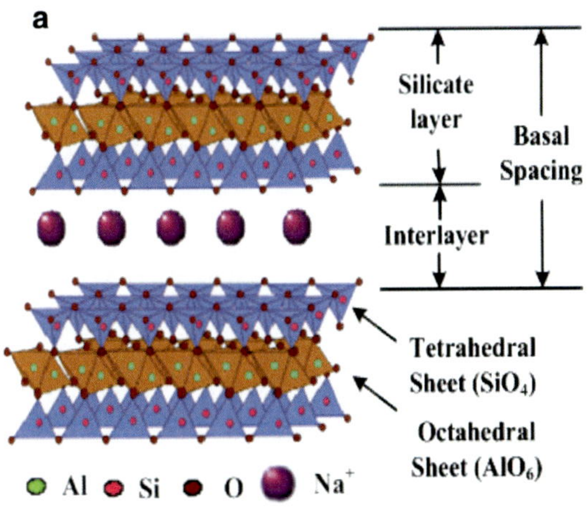


Figure 1. (a) Molecular structure of MMT containing exchangeable sodium ion (MMT-Na+), [KIM06, reprinted with permission from Elsevier].

$$RH^+ + M^+\text{-Clay} \leftrightarrows RH^+\text{-clay} + M^+ \text{ [RUI04]}$$

With R being an organic base capable of protonation, $M^+$, the exchangeable inorganic cation.

The success of the cationic exchange depends on several parameters like the mineral host structure and the nature of the compensating cation.

### 6.2.1. Influence of the Mineral Host Structure

The excess negative charge of layered silicates and their capability to exchange ions can be quantified by a specific property known as the cation-exchange capacity (CEC) and expressed in milliequivalent per gram (meq/g). In other words, the CEC corresponds to the number of monovalent cations it is possible to substitute to the sodium and/or calcium initially present at the platelet surface in order to balance the charge deficit of 100 grams of clay. The CEC of a montmorillonite is usually included between 60 and 150 meq/100 grams. So, in a cationic exchange, the quantity of organic ions increases with the clay charge density. But, in the case of micas, this quantity remains low despite a high charge density. Because, the tetrahedral location of charge leads to strong interactions between the layers and prevents the hydratation of cations, the galleries opening and consequently reduces the capacity for cationic exchange [LUC99]. In the case of vermiculites, the compensating cations can be hydrated because of a lower charge density. But the vermiculite CEC is very high (about 200 meq/100g), a too important cationic exchange can be harmful for monomer or polymer diffusion that will be limited by a too strong sterical hindrance [LAN95]. The isomorphous substitutions of montmorillonite take place in octahedral layers. The electrostatic interactions with the organic cations are attenuated by the tetrahedral layer. The montmorillonite (and also hectorite) show the most interesting compromise between a high enough CEC to get a efficient cationic exchange without sterically congesting the interlayer gallery and a low enough CEC to allow the layers separation in aqueous medium.

### 6.2.2. Influence of the Compensating Cation

In order to obtain the exchange of the organic-base salts with the cations in the galleries, water swelling of the silicate is needed. And the nature of cations has an influence on the silicate swelling. For this reason, alkali cations are preferred in the galleries because 2-valent and higher valent cations prevent swelling by water and are less exchangeable. Indeed, the hydrate formation of monovalent intergallery cations is the driving force for water swelling. Natural clays may contain divalent cations such as calcium and require exchange procedures with sodium prior to cationic exchange [MAC59]. More mobile and smaller the cation is, easier is the cationic exchange. The more usual compensating cations in clays can be classify by increasing order to be cationically exchangeable :

$$Cs^{+} < Rb^{+} < Ca^{2+} < Mg^{2+} < Na^{+} < Li^{+}.$$

## 6.3. Choice of Organic Cation

Since the first works of Gieseking [GIE39] and Hendricks [HEN41], numerous organic salts of clay minerals have been prepared and studied to make organophilic clays and introduce them into polymer matrix. In ninety years, the first surfactants used were amino acids that were successfully used in the synthesis of polyamide 6-clay nanocomposites [OKA90]. But, in most cases, the organic bases are protonated aliphatic or aromatic amines [LEP02, BUR03]. Recently, nitrogenated or phosphonated heterocycles are used. The organic cations can have several key roles. They are compatibilizers by lowering the surface energy of the silicate surface and by improving wetting with the polymer matrix [BOU02, LEP05]. They can be a dispersion aid by increasing the gallery height thanks to their long organic chains tethered to the silicate surface. The polymers can diffuse between the layers and eventually separate them [HER07,VER05, VER07]. Sometimes, the organic cations may even provide functional groups that can react with the polymer [NGO09] or initiate polymerization of monomers [CHI01]. If the organic cation bears more than one function (e.g. cationic fluorescent dyes contain charge-bearing and fluorophore-bearing groups), cationic exchange process is still possible and suitable to functionalize clays, for example by tailoring their optical properties [CAP04]. Conclusively, the cationic exchange surface modification tailors the interfacial properties between the hydrophilic clay and the hydrophobic polymer. It must be a help to exfoliate the clay layers and ensures a interactions continuum between nanofillers and matrix.

### 6.3.1. Alkylammonium Ions

The alkylammonium ions are the most used ionic liquids to modify clay. In the synthesis of nanocomposites, the main role of alkylammonium ions is to lower the surface energy of the layered silicate in order to reduce the electrostatic interactions between the silicate layers and allow molecules to diffuse between the layers. The most widely used alkylammonium ions are based on primary alkylamines put in an acidic medium to protonate the amine function. Their basic formula is $CH_3-(CH_2)_n-NH_3^+$ with n varying from 7 to 17. The chemical structure has a significant impact on cationic exchange. The carbon chain length, the size and the form of the polar head and the organic groups carried by the ion have a strong influence on the success of the cationic exchange.

- The increase of the interlayer spacing is dependent on the ion alkyl chain length [PIN00, LANG4, ZIL99]. Thus, by increasing the alkyl chain length, the entropic contribution of the adsorption energy is increased. As a result, more Van der Waals (VdW) interactions are developed. However, the adsorption is carried on beyond the clay CEC via VdW interactions between alkyl chains for ions with more ten carbon atoms [JOR49-COW58]. This excess of alkylamonium ions leads to the new layers adsorbed onto the platelets surface giving back partially its hydrophilic character to the clay due to the negative conterions adsorption [PAT93]. Then, the increase of the alkyl chain length leads to the decrease of the critical micellar concentration for the same initial molar percentage. Increasing the temperature is then necessary to well disperse the alkylammonium ions and succeed the cationic exchange.
- Fixation is favored if the organic cations contain radicals capable of interacting by H-bonding with the surface oxygens of the silicate. Since these are arranged in ditrigonal six-membered rings, the presence of groups with trigonal symmetry, *i.e.*, -$NH_3^+$ groups in alkylammonium ions, favors reaction. The exchange is also favored if the ion owns aromatic rings capable to exchange Π electrons with the surface. The methylene blue is well-known to be easily adsorbed on clay platelets and is typically used in the surface energy determination.
- The tethering of organic cations also depends on size and form of polar head [ROW63]. The small cations like $-NH_4^+$ and those having one or two methyl groups can be located into the surface hexagonal cavities and increase the bonding strength with the mineral host structure. The bonding strength decreases from primary, to secondary and finally tertiary amines. Moreover, this reinforcement of the ion/clay interactions decreases the reversibility of the cationic exchange. But the quaternary amines have a specific behaviour similar to the primary and secondary amines ones. The works of Mc Atee [MAC59] showed that the adsorption of the ions derived from primary amines was not sufficiently to reach the CEC unlike to quaternary ammoniums.

The main ammonium ions that are used to modify the lamellar silicates and that are available at different suppliers are described in Table 1. The clay surface properties can be tuned as a function of the organic groups bonded to the ammonium ion. The clay can be quite dispersive and show a surface energy similar to polyolefin one when the modifying ion carry two tallow chains of about 18 carbon atoms or one tallow chain and an aromatic group. The clay can be more polar when it is modified by short chains or a carboxylic function. At last, montmorillonite can have an intermediate character, both polar and dispersive when the lamellar silicate is functionalized with a quaternary ammonium showing two alcohol groups and one tallow chain.

**Table 1. Different ammonium ions used to modify the commercial montmorillonites [BUR03]**

| Chemical structure | Trade name | Supplier | $\gamma_s$ ($mJ.m^{-2}$) |
|---|---|---|---|
| $CH_3$-$(CH_2)_n$-$NH_3^+$ With 7<n<17 | - | [1]Southern Clays (USA) | $40 > \gamma_s > 34$ |
| 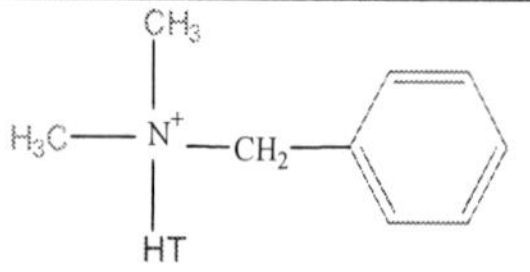 | [1]Cloisite® 10A<br>[2]Nanofil® 919<br>[4]Tixogel®MP250 | [2]Süd Chemie (GERMANY) | 30.0 ± 2.0 |
| 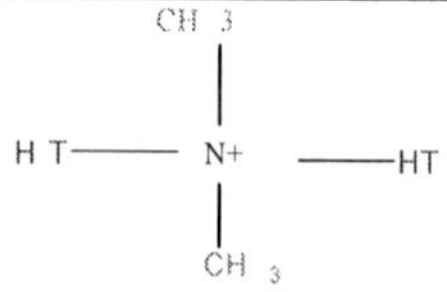 | [1]Cloisite® 15A<br>[2]Nanofil 15<br>[2]Nanofil 948<br>[3]Lucentite®SAN<br>[4]Tixogel®MP100<br>Nanocore I44 | [3]Unicoop Chemical (JAPAN) | 25.4± 1.0 |
| 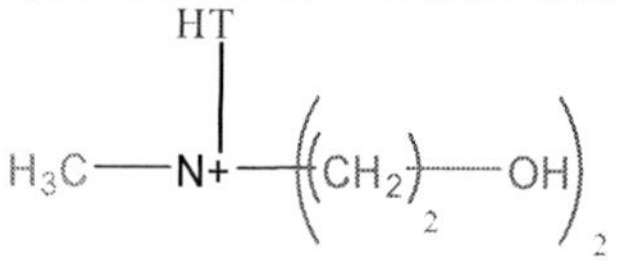 | [1]Cloisite® 30B<br>[2]Nanofil® 804 | [4]Société Française Des Bentonites | 34.5 ± 2.0 |
| 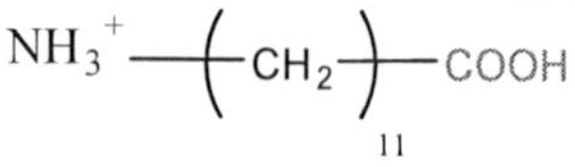 | [2]Nanofil® 784 | [5]Nanocor | 38.0 ± 1.0 |

Note: HT means hydrogenated tallow which is a fatty chain distribution with the following composition : 65%$C_{18}$, 30%$C_{16}$ and 5%$C_{14}$.

### 6.3.2. Aminoacid Ions

Amino acids are molecules, which consist in a basic amino group (-$NH_2$) and an acidic carboxyl group (-COOH). In an acidic medium, a proton is transferred from the –COOH group to the intramolecular –$NH_2$ group. A cationic exchange is then possible of -$NH_3^+$ function formed for an inorganic cation (i.e. Na+, K+...) intercalated between the silicate layers so that the layered silicate is rendered organophilic. A wide range of ω-amino acids ($H_3N^+$-$(CH_2)_{n-1}$COOH) have been intercalated between the layers of montmorillonite [USU93-KOR01]. Amino acids were successfully used in the synthesis of polyamide 6-clay hybrids [USU93] because their acid function has the ability to polymerise with ε-caprolactam intercalated between the layers. Thus, this intragallery polymerisation exfoliates the clay in the polymer matrix and a nanocomposite forms.

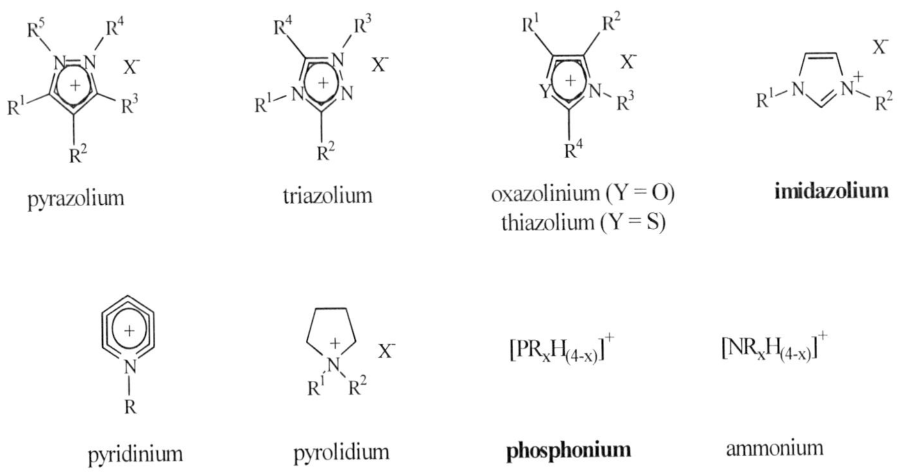


Figure 2. Different families of ionic liquids used as surfactants.

## 6.3.3. Thermostable Ionic Liquids

Ionic liquids (ILs) are organic salts with melting points below 100°C. Ionic liquids can be produced with a variety of cations and anions. The most common organic cations containing nitrogen are imidazolium and pyridinium derivatives; other compounds are based on phosphonium or tetra-alkylammonium cations while commonly used anions include [BF4]-, [PF6$^-$], [CF3COO$^-$], and [CF3SO3$^-$], etc. In addition to the ability to control the hydrophilic/lipophilic balance of the IL structure through proper selection of the constituent ions, an important property of ILs is their high thermal stability.

This can be used to advantage to overcome the lower thermal stability of conventional organic modifiers (*i.e.* alkylammonium ions) that are used to enhance dispersion, wetting and compatibility of nanoparticles in organic media. Thus, it is not surprising that the bulk of the recent literature focuses on polymer nanocomposites as the intended application of IL modified nanomaterials. Treatment of layered silicates, carbon nanotubes or other nanoparticles with appropriate ILs may open new avenues for the preparation of polymer nanocomposites from high temperature thermoplastics or from thermosets with high cure temperatures. Important incentives were the improvement of the thermal stability of the organoclays and the preparation of nanocomposites with enhanced flame retardant characteristics. Efforts have been made to synthesize thermally stable organoclays based on imidazolium [Bou03], pyridinium or phosphonium surfactants [XIE02],[CAL08]. Figure 2 displays the whole of ionic liquids used as surfactant for clay. The utilization of these new ionic liquids as the surfactant increases the initial decomposition temperature [GIL00] and improves the thermal stability [XIE02-MI07] as shown in Figure 3.

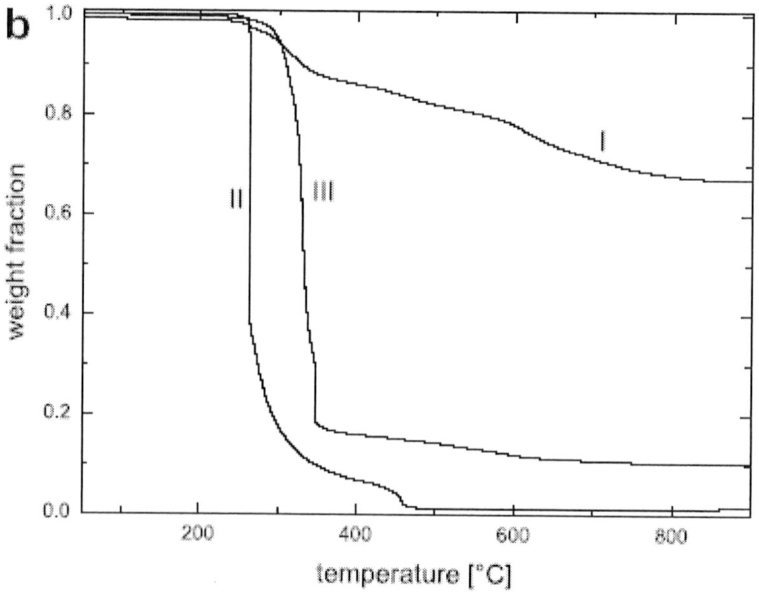


Figure 3. Comparison of the TGA thermograms of pure PP (II), imidazolium clay (I) and the nanocomposite containing 4 vol% of MMT modified with imidazolium (III). [MIT07, reprinted with permission from Elsevier]

## 6.4. Consequences of the Cationic Exchange on Lamellar Silicates

The cationic exchange usually takes place in water or in a mixture of water and alcohol or in an organic solvent. The use of permuted water is always worthy to be preferred in order to get rid of any desired cation and to make the clay swelling easier. But sometimes the addition of a co-solvent is necessary to completely dissolve the organic modifying ions. First, the lamellar silicates are dispersed in the exchanging medium under stirring. Separately, the modifying ions are dissolved in the required amount of co-solvent and the obtained solution is added to the clay suspension. Whatever the nature of the adsorbate (organic ions) and regardless of the driving forces which make it possible, diffusion is critical because it influences molecular mobility. Consequently, the efficiency of the cationic exchange process depends on several parameters like temperature, reaction time and modifier concentration. Clay cationic exchange involves indeed electrostatic interactions (compensation of electrical charges) but also weak intermolecular interactions (physisorption) between the adsorbent (clay platelet surface) and the adsorbate. The adsorbent surface may be either the external surface of clay platelets, or the internal surface of clay galleries, or even the border of clay platelets-depending on the electrical charges on the surface of the layers, the nature of the interlayer compensating cations and the sterical hindrance effect due to the difference between the adsorbate molecular size and the pristine $d_{001}$spacing of the clay. Even if a small part of the charge balancing cations is located on the external crystallite surface, the majority of these exchangeable cations are located inside the galleries. On the basis of the CEC of the clay, the content of the surfactant is usually about 35-45 wt% [LEP02, LEP03, BUR04].

### 6.1.1. On the Organization of the Organic Chains in Clay Gallery

*Effect of alkyl chain length*

In general, the longer the surfactant chain length, and the higher the charge density of the clay, the further apart the clay layers will be forced. This is expected since both of these parameters contribute to increasing the volume occupied by the intragallery surfactant [LEB99]. But the interlayer distance also depends on the way of the onium ions chains organize themselves in the gallery. The arrangement of organic cations within the interlayer space depends, essentially, on the three factors : the size of the organic cation, the charge density and the location of charge (tetrahedral or octahedral) in the silicate layers. The organization of long-chain alkylammonium cations in the interlayer space of 2:1 clay minerals has been studied extensively [LAG86, OGA97]. With montmorillonite, where the charge density is low and predominantly octahedral, organic cations lie parallel to the silicate layers with an all-trans conformation. But when the alkyl chain length is varied from octyl, to octadecyl, the chain conformation evolves from monolayer ($d_{001}$ = 13.5 Å) to bilayers ($d_{001}$ = 17.4 Å) up to paraffin-type structures ($d_{001}$ = 31 Å). Conformational changes are possible through rotations around C-C bonds (kinks). This occurs for pseudo-trimolecular arrangements where some chains ends are shifted above one another and the spacing (ca 22 Å) becomes the thickness of three alkyl chains. As shown by the data in Table 2, basal spacings measured by XRD increase in proportion to the alkyl chain length of the ammonium cations.

To evaluate the chains conformation in the clay galleries, the theoretical length is estimated when the chain adopts an all-trans conformation. The theoretical extended chain length, $l_t$, can be calculated as follows:

$$l_t\,(\text{Å}) = 1.265\,(n_c-1) + 3.0$$

where ($n_c$-1) is the number of methylene groups in the alkyl chain, 1.265 Å is the contribution due to the $-CH_2-$ chain segments by assuming that the chains adopt an all-trans configuration and the dimension of the methyl end group (3 Å) is added as proposed by Lagaly [LAG86].

**Table 2. Interlayer distance ($d_{001}$, Å) of alkylammonium exchanged montmorillonites [LEP03, reprinted with permission from Wiley].**

| Sample Name | Alkylammonium ion | $d_{001}$ spacing (Å) [a] | Calculated value, $l_t$ (Å) | Tilt angle α (°) |
|---|---|---|---|---|
| MMt-$Na^+$ | none | 12.6 | - | - |
| MMT-C18 | $CH_3$-$(CH_2)_{17}$-$NH_3^+$ | 31.6 | 24.5 | 50 |
| MMT-C16 | $CH_3$-$(CH_2)_{15}$-$NH_3^+$ | 29.6 | 22 | 50 |
| MMT-C12 | $CH_3$-$(CH_2)_{11}$-$NH_3^+$ | 22.3 | 18.2 | 32 |
| MMT-C8 | $CH_3$-$(CH_2)_7$-$NH_3^+$ | 13.6 | 11.9 | 4 |

[a] Measured by XRD at 0.1 °/min

Vaia et al. give a more realistic description by monitoring frequency shifts of the asymmetric $CH_2$ stretching and bending vibrations. They show that alkyl chains can vary from liquid-like to solid-like, with the liquid-like structure dominating as the interlayer density or chain length decreases, or as the temperature increases.Thus, the chains with eighteen carbon atoms show a crystal liquid behaviour whereas the chains with less twelve carbon atoms have a liquid like behaviour and nearly a gas liquid like behaviour from six carbon atoms. Thus an order-disorder thermal transition can be measured for octadecylammonium exchanged montmorillonite (about 73°C) whereas the octylammonium exchanged montmorillonite does not show thermal transition [LEP03]. Since a few years, the molecular dynamics simulations confirmed the complexity of organizations adopted by alkyl chains into the interlayer spacing. According to Hackett et al [HAC98], for the same CEC and the same interlayer distance, when the alkylchain length increases, the pressure into the gallery increases and drives the packing of the chains up to the pressure exceeds the cohesion strength of the layers and leads to the expansion of the interlayer distance.

*Effect of temperature*

Temperature at which is performed the cationic exchange has a clear effect on the layer structure in the clay gallery. In the Le Pluart works, the authors show that the octadecylammonim chains arrangement in the gallery is clearly different as a function of the cationic exchange temperature (see Figure 4). The montmorillonite modified at 60°C displays on the X-ray spectrum two quite brad peaks, which means that two organizations coexist in the galleries. The first peak ($d_{001}$ = 33.4Å) is related to intercalated ions with a paraffin-like structure whereas the second one ($d_{001}$ = 20.2Å) is rather significant of a pseudotrilayer structure. On the other hand, when the cationic exchange is performed at 80°C, the X-ray diffraction spectrum realized on modified clay presents two peaks as well but they are significant of only one single organization since the peak at 16Å is a second-order peak of the first peak at 32Å. Moreover, these peaks are much narrower than those of montmorillonite modified at 60°C, which reveals a very stable and very well-defined paraffinic structure.

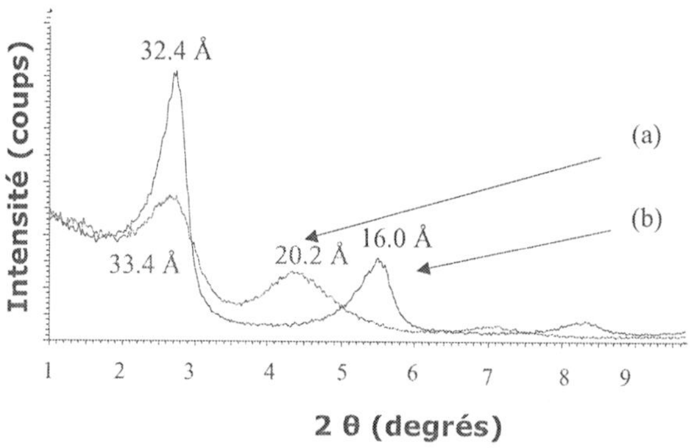


Figure 4. XRD spectra (from 2θ = 0 to 10°) of montmorillonite modified with octadecylammonium ions with 1 CEC a) @60°C and b) @80°C [LEP02].

*Effect of amine/clay ratio*

The amine/clay ratio is another parameter that governs the cationic exchange of montmorillonite [POS01]. An excess of amine is necessary to complete the layer of ions ionically bonded to the platelet surface. Usually, the cationic exchange of the octadecylammonium ions is carried out at 80°C with an excess of amine corresponding to 2CEC. The molecular modelling showed by the minimum potential energy calculation that the chain conformation evolves from a monolayer to a bilayer structure by doubling the organic cations amount introduced.

## 6.4.2. On the Interactions between Clay/Modifying Ions

Thermogravimetric Analysis (TGA) measurements may help completing characterizations of interactions between clay and surfactant. The degradation temperatures corresponding to weight losses between 150°C and 250°C (up to 300°C) typically indicate the presence of organics which could be simply physisorbed (or eventually interacting by hydrogen bonding) either on the external surfaces of the silicate layers or at their edges. Such organic fraction either didn't undergo the cation exchange process, or stayed unconfined, or maybe entered the galleries but remained in a peripheral position. On the other hand, weight losses in between 300°C and 550°C rather correspond to organics intercalated into clay galleries [LEP03]. Indeed, organics inside clay galleries are better insulated and thus protected from thermal degradation; in addition, the eventual products of an initial thermal decomposition are slowed down from diffusing to the gas phase by the presence of the silicate layers (barrier effect). A suited washing procedure is expected to reduce the peaks corresponding to any chemical physisorbed at the edges or on the external surfaces of the silicate layers, still preserving the peaks which denote the chemicals intercalated into clay galleries [LEP03]. On the TGA thermograms reported in Figure 5, the two weight losses due to the physisorbed and intercalated ions are clearly observable. Repetitive washings with a hot ethanol:water mixture lead to a complete disappearance of the physisorbed ions without removing the ionically bonded ions in the galleries.

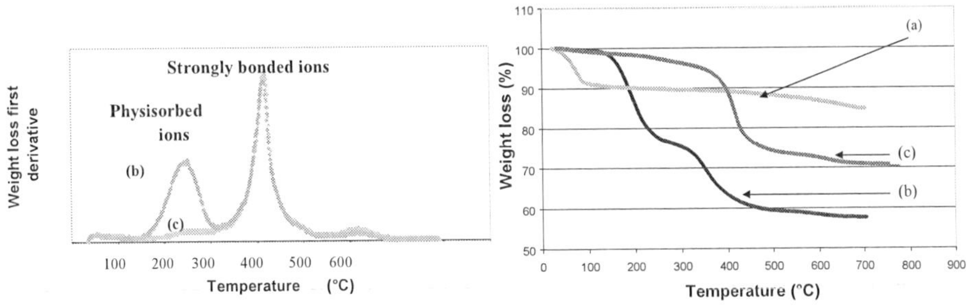


Figure 5. TGA experiments on the pristine sodic clay (a), after exchange before washing (b) and after repetitive washings (c): weight loss and weight loss first derivative as a function of temperature [LEP03, reprinted with permission from Wiley].

# 6.5. Reactivity of Lamellar Silicates

The increase of the interfacial surface inherent to the nanometric fillers disturbs the environment of macromolecular chains around the nanofillers up to 50 nm from the surface. The properties of polymer matrix are consequently modified close to the surface by modification of reactions taking place near the surface respect to ones in bulk and by decreasing chain mobility.

### *Effect on copolymerisation*

Nanofillers by their hydroxyl groups on the layer edges and the functionalities carried out by the modifying ions may strongly participate to the chemical reaction and may copolymerise with the reactive medium. Some cationic modifiers intercalated into the clay gallery can initiate the polymerization of monomer. As mentioned above, the organophilic montmorillonite ion-exchanged with 12-aminododecanoic acid allowed the polymerization of ε-caprolactam monomer to prepare the nylon –clay nanocomposite [OGA97]. In the formulation of the epoxy-amine adhesive [BUR03], the clay addition into the amine monomer is judicious to help the interactions between the montmorillonite functionalized carboxylic acid and amine. The CP-MAS $^{13}C$ NMR spectrum shows a very fine resonance about 177 ppm which seems significant of a carbonyl group characteristic of an amide group. There is reaction at the surface of layers between the carboxylic group located on the layers and the amine monomer to form an amide group. After polymerisation, the epoxy-amine network is intimately bonded to the lamellar silicates. Another example is when the lamellar silicates are exchanged with 2-methacryloyloxyethyl hexadecyldimethyl ammonium bromide surfactant (denoted as MHAB), the silicate surface is made not only organophilic but also reactive thanks to methacrylate group by which copolymerization with vinyl monomers may be possible. Thus, for the preparation of polystyrene nanocomposite, the styrene monomer inserted into the interlayer gallery may react with the MHAB double bond and synthesize *in situ* a MHAB-PS copolymer. This copolymer grafted covalently onto the silicate surface will be miscible with the polystyrene chains of the matrix and will generate a strong interphase with PS matrix [NGO09].

### *Effect on polymerisation kinetics*

All the authors agree on the fact that organophilic clay catalyzes the epoxy reaction. Le Pluart et al. [LEP05] show that the homopolymerization of DGEBA alone is favored by organophilic lamellar silicates and decrease the temperature of beginning the reaction. Similar effects were noticed by Butzloff et al [BUT01], Ingram et al.[ING08], Chin et al. [CHI01] and Park et al. [PAR03]. The explanation is that the clay initiates an electrophilic attack on the epoxy ring and thus initiates an anionic ring opening polymerization to form an ether- linked polymer. Even with complete systems, with hardener, the lamellar silicates may have a catalytic effect on polymerisation reactions. Monitoring of the viscoelastic properties of DGEBA/MCDEA based systems with a multifrequency test during an isothermal curing at 135°C reveals that gel time of this epoxy/amine system is decreased in presence of organoclay (Figure 6). In order to highlight this, only two of the five frequencies used during the multiwave characterization are plotted. When these gel times are reported on the curves of the evolution of conversion vs. isothermal time of curing (Figure 7), one can notice that the conversion at gel is not significantly modified by the presence of organoclay and corresponds to the value of 0,58 that can be calculated from Macosko-Miller theory. These two observations lead to the conclusion that the organoclay does not modify strongly the reaction taking place between the epoxy and the amine, but only accelerates it [LEP05].

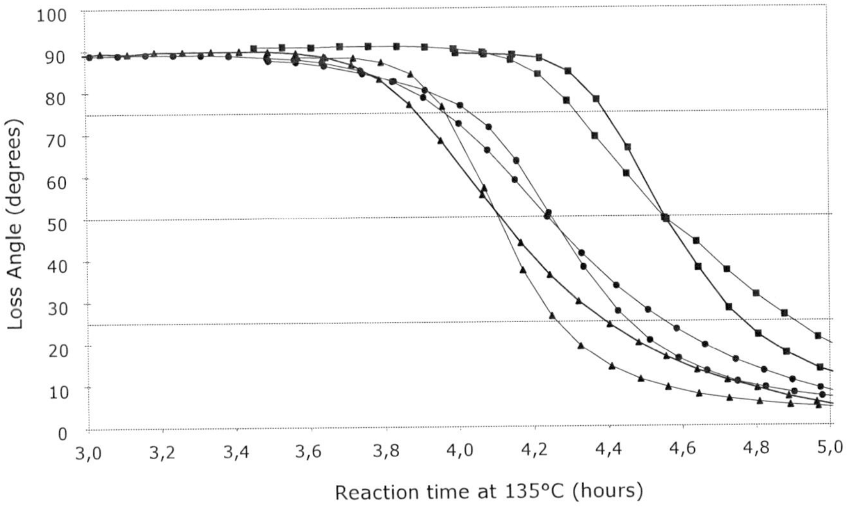


Figure 6. Loss angle evolution vs time of isothermal curing at 135°C (frequencies 10 and 100 rad.s-1) for neat DGEBA/MCDEA (squares), +5 phr TixogelMP250 (circles) and + 5 phr octadecylammonium exchanged montmorillonite (triangles) [LEP05].

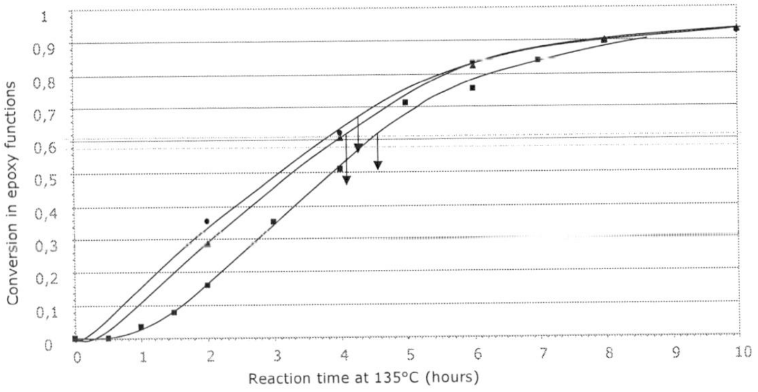


Figure 7. Epoxy functions conversion vs isothermal time of curing at 135°C for neat DGEBA/MCDEA (squares), +5 phr TixogelMP250 (circles) and + 5 phr octadecylammonium exchanged montmorillonite (triangles) [LEP05].

The catalytic power of the organoclays is attributed to the OH groups present on the layers edges. By using a montmorillonite exchanged with carboxylic acid (MMT-COOH), the opening of epoxy rings was checked by $C^{13}$NMR. The Works of Wang et al explain this catalytic effect by the acidity of primary ammonium inserted into the clay gallery [WAN94]. The gelation time given in Table 3 is shorter when the MMT-COOH is first dispersed in the epoxide monomer than introduced into the amine monomer. It seems that the strong interactions between modified montmorillonite and amine must stabilize, even immobilize a part of amine monomer around the layers. Thus, the amine is less available to react with the epoxide comonomer, which can explain the reaction slows down.

**Table 3. Gelification times (@60°C) for an epoxide-amine network without clay, with 5 pcr of carboxylic acid functionalized montmorillonite first introduced in epoxide or in amine monomers**

| | Without clay | With MMT-COOH introduced in Epoxyde | With MMT-COOH introduced in Amine |
|---|---|---|---|
| Gelation time (min) | 28 | 18 | 35 |

## 6.6. Photo-Functionalization of Lamellar Silicates

***Cation exchange process*** is potentially adequate to confer specific (*e.g.* optical) properties to any layered (natural or synthetic) mineral filler able to exchange cations by dispersing in solutions containing water and/or an organic solvents [ENDO86, KAN04, SAS05]. Bujdak [BUJ06] has recently analysed and critically reviewed most of the older papers dealing with the photo-functionalization of clay minerals and their interactions with cationic organic dyes. López Arbeloa *et al.* [LOP07] focused their review on the interactions between rhodamine dyes and clay layered films. *MMT* is the most common lamellar mineral filler chosen by researchers to study its interactions with rhodamines [SAS05, LOP07] and other organic dyes [KAN04, OGA03], but other minerals have been evaluated (fluor-taeniolite [FUJ97], laponite [SAL08], synthetic saponite [IWA00], sepiolite [RYT00], fluorohectorite [CZI04], synthetic mica [MIY00], zeolite [VIE98], kaolinite [MIY00]) as well as other cationic organic dyes (1,1'-diethyl-2,2'-cyanine [IWA00, MIY00], oxazine-4 [KAN04], azobenzene [OGA03], methyl green [RYT00], methylene blue [CZI04], acridine orange [GAR97], thiamine [CZI04]). *Fluorescent molecules* are fickle but versatile tracers, sensitive to any difference in the surrounding environment (*chemical composition, atomic arrangement, molecular configuration, physical confinement, temperature*). Against any preconceived opinion, the use of fluorescent molecules isn't only a prerogative of biologists : fluorescent tracers and dyes have already been successfully used in the field of polymer processing – to monitor shear effects on phase segregation in the mixture of two miscible polymer by *in situ fluorescence quenching* [MAN91], to evaluate the mixing efficiency of an internal batch mixer [CAS03], to estimate the RTD during extrusion [CAS05], to measure the temperature of molten polymers during extrusion and injection moulding [BUR04a], for real-time process monitoring of biaxially stretched polypropylene films [BUR04b] and, recently, to assess the size and distribution of fillers in a polymer matrix [ALO06]. Considering the number of available on-line monitoring techniques based on some optical property or detection system and the multiplicity of diversified applications, fluorescence offers huge potentialities in more than a research field.

### 6.6.1. Spectrofluorimetry

***Fluorescence*** is a three-stage process that generally occurs in polyaromatic hydrocarbons or heterocyclic molecules called *fluorophores* or *fluorescent dyes*. In the first stage, a photon of a given energy is supplied by an external source (*e.g.* an incandescent lamp or a laser) and absorbed by the fluorescent molecule, creating an *excited electronic singlet state*. This excited state exists for a finite time, during which the molecule undergoes conformational changes and is subject to several interactions with its molecular environment. Such processes partially dissipate the energy of the excited state, yielding a *relaxed singlet excited state* and thus producing a fluorescence emission. Finally, a photon is emitted returning the molecule to its ground state. Due to the energy dissipations in the excited-state lifetime, the emitted photon has a lower energy (longer wavelength) than the absorbed one: the difference in energy (or wavelength) of the two photons is called the *Stokes shift*. The sensitivity of any fluorescence technique strictly depends on this fundamental parameter, which allows emission photons to be detected against a low background, clearly isolated from excitation photons. In theory, fluorescence is a cyclical process: unless the fluorophore is irreversibly destroyed in the excited state (phenomenon known as *photobleaching*), it can be repeatedly excited and detected.

Fluorophores can assemble in *dimers* and/or *higher-order aggregates*: any possible configuration produces a well-defined absorption spectrum and doesn't assure the same fluorescence emission. *J-aggregates* (head-to-tail molecular arrangements) typically exhibit a narrow absorption peak (the so-called *J-band*) which is red-shifted with respect to the monomer band (*bathochromic shift*), whereas *H-aggregates* (tail-to-tail molecular arrangements) show a broader absorption peak (*H-band*) which is rather blue-shifted with respect to the monomer band (*hypsochromic shift*). On the basis of the exciton theory, J-aggregates are fluorescent whereas H-aggregates are affected by self quenching. Interactions, either between two fluorophores or between a fluorophore and the other species in the surrounding environment, can produce environment-sensitive fluorescence emission.

***Fluorescence quenching*** is a bimolecular process which reduces (and sometime zeroes) the fluorescence signal intensity without reshaping the emission spectrum: it can result from transient excited-state interactions (collisional quenching) or from the formation of non-fluorescent ground-state species (*i.e.* H-aggregates). Quenching tends to occur when high fluorophore concentrations are used. Spectrofluorimetry is an essential tool to check the result of a photo-functionalization process.

### 6.6.2. Fluorescent Ions

***Fluorescent probes*** absorb the radiation at a characteristic wavelength and then emit a fraction of the absorbed energy at a longer wavelength. If the fluorescent tracer is adequately diluted, the intensity of its emission is proportional to the incident luminous power P0 as well as to the local concentration of tracer *C* and to the concentration *C* averaged on the whole optical pathway throughout the solution. The chemical structures of the most used dyes are reported in Figure 8.

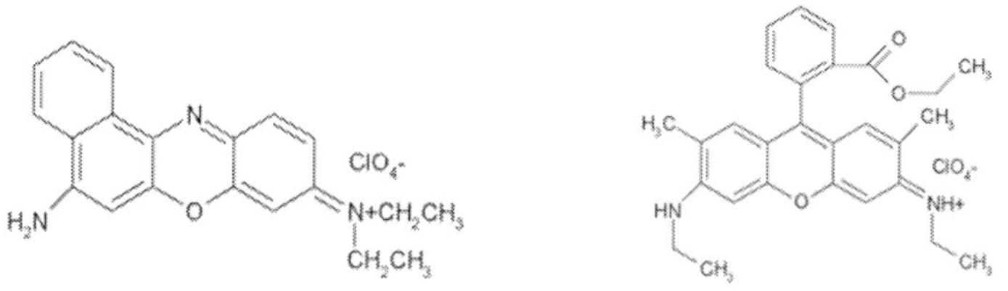


(a) Nile Blue A perchlorate (b) Rhodamine 6G Perchlorate

Figure 8. Chemical structure of the most used fluorescent dyes.

- ***Rhodamine*** *dyes* are the most frequently used class of fluorescent tracers: they are typically excited by a radiation having a wavelength of 470-520 nm and they emit at a wavelength comprised between 550 and 600 nm. *Rhodamines* have been used since a long time [ENDO86] and are nowadays quite familiar to many research teams. Indeed, they represent the *ideal probes to study heterogeneous systems* thanks to the strong dependence of their absorption and fluorescence emission on the properties of the host matrix: their optical properties are *environmentally sensitive* and according to the theory of exciton splitting depend on the arrangement and the spatial configuration of the dye molecules (*monomers*, *dimers*, *J-aggregates*, *H-aggregates*) [CAP04]. The organization of the photoactive species within clay inorganic microstructure depends on the *host-guest* and *guest-guest interactions*: the adsorption of rhodamine onto clays sometimes leads to *metachromasy* in the absorption spectrum (*i.e.* shift of the main absorption band) [CAP04].

- ***Nile Blue*** has a polyaromatic heterocyclic configuration based on the benzophenoxazine. Its wavelength range of absorption is about 627 nm and its fluorescence emission situated about 665 nm shows clear confinement effects on the amplitude of the Stokes shift and the dynamics of solvation in ethanol, when confined to sol-gel glasses with 50 Å and 75 Å average pore size. In particular, its fluorescence emission undergoes a blue shift for confinements inferior to 7.5 nm [BAU01]. If nano-confined in 5 nm pores of the same sol-gel glass but in a non-polar medium such as dodecane, Nile Blue shows an even more pronounced blue shift, from 664 (ethanol) to 648 nm (dodecane) [MAU04]. Nile Blue A Perchlorate (NBAP) has already been used for cationic exchange processes on commercial clays [MAU04, GIL04].

Cationic fluorescent dyes are easily adsorbed on clay mineral surfaces by cationic exchange process: previous works have shown that the intensity of their fluorescence emission depends on both the nature of the adsorbent clay and the concentration of the adsorbate dye – the latter being correlated to the *Cationic Exchange Capacity* (CEC) of the clay. Natural clay can contain significant amounts of *Fe in the octahedral layers*: some authors assert that the presence of iron can cause fluorescence quenching – but quenching can also occur at high levels of adsorbed dye (*concentration quenching*) because of dye-dye interactions, as shown in Figure 9 [MAU04].

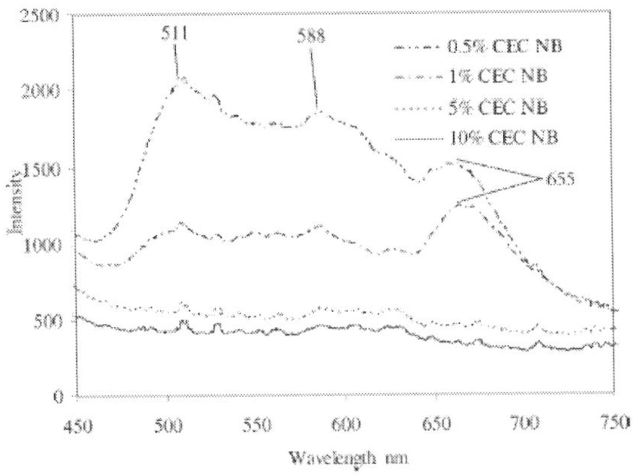


Figure 9. Fluorescence spectra of an organically modified clay containing different amounts of Nile Blue (excitation wavelength 407 nm, integration time 4s) [MAU04, reprinted with permission from Wiley].

### 6.6.3. Photo-Functionalization by Cationic Exchange

The cationic exchange process traditionally employed has been inspired by some recent works of Maupin, Bur, Gilman *et al.*, a group of researchers at the National Institute of Standards and Technology at Gaithersburg, Maryland, USA [MAU04, GIL04]. Since ion exchange is a reversible process, clay functionalization is generally performed in the presence of an excess of surfactant – typically twice the value of the cation exchange capacity of the clay (2 CEC) – in order to force the reaction in the desired direction. However, when the adsorbate is a photo-active chemical and the ion exchange is aimed to the photo-functionalization of the adsorbent, a stricter control on the concentration of surfactant is required in order to avoid saturation (*i.e.* fluorescence quenching). Moreover, the use of an organically-modified clay which has already undergone a first cationic exchange process is often required: we'd better talk of modifier concentration (or residual cationic exchange capacity) instead of cationic exchange capacity *tout court*. This means that most of the $Na^+$ sites originally available are likely to be no longer exchangeable. On the other hand, it is highly probable that a certain (significant) amount of such sites corresponding to the residual CEC is still represented by inorganic cations, for the organic molecules are surely big enough to be an obstacle to each other because of sterical hindrance. Since the residual value of CEC for organophilic clay is less than the original value for the inorganic clay, the introduction of an amount of fluorescent molecule equivalent to the residual value of CEC into the exchanging medium would be already enough to be in the presence of an excess. In the works of Esposito [ESP09], the cationic exchange processing is well described. The organophilic clay has been cationic-exchanged with two cationic organic dyes (Nile Blue A Perchlorate and Rhodamine 6G Perchlorate), introduced in two different concentrations (corresponding to 1 and 0.25 times the initial modifier concentration of the organically modified clay, denoted 1MC and 0.25MC respectively) in a solution 90/10 permuted water/ethanol. The use of permuted water is always worthy to be preferred in order to get rid of any undesired cation. The correct amount of ethanol to be introduced in the exchanging medium has been determined as the smallest amount of co-solvent necessary to completely dissolve the fluorescent molecule, *i.e.* in order to get a saturated solution. First, the permuted water has been warmed up to 80°C. Then, the organically modified clay has been added to the stirring hot water and let disperse for at least 15 minutes. Separately, the fluorescent molecule has been dissolved in the required amount of ethanol and the obtained solution has been added to the clay suspension. Once started, the cationic exchange process has been performed under mechanical stirring for 24 hours. After additional 24 hours in the still exchanging medium, the photo-functionalized clay has been separated from the liquid by centrifugation (20 min at 4000 rpm), washed with ethanol and recovered anew by centrifugation. The described washing procedure has been repeated until a sufficiently colourless supernatant was obtained (meaning that no more fluorescent molecule could be extracted by washing), then let dry under exhaust hood for several days at room temperature.

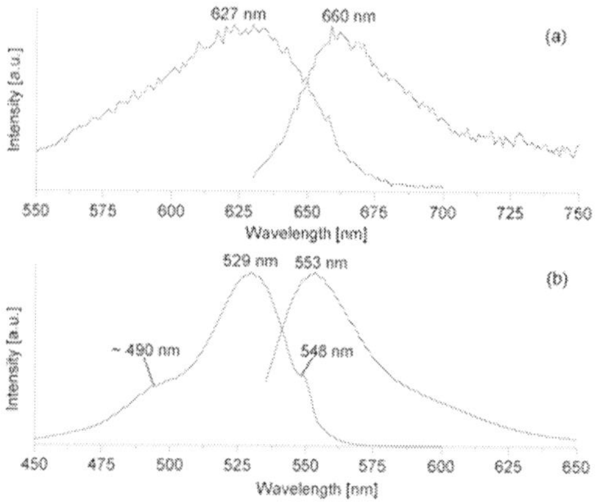


Figure 10. Absorption and emission spectra in ethanol of C30B clay photo-functionalized by cationic exchange with (a) NBAP 1MC and (b) RhP 1MC. Samples have been carefully washed with ethanol in order to assure that the response has to be assigned uniquely to the intercalated dye. [ESP08].

If we take a look at the results obtained by spectrofluorimetry on the washed photo-functionalized samples, we realize that the photo-functionalization was successful with both the fluorescent molecules (Figure 10). In terms of absorption: NBAP-clay absorbs in the same range as the raw cationic dyes (no significant metachromasy) while RhP-clay, as expected, shows a slight metachromatic effect (the main absorption band is blue-shifted, towards higher energies). Vice versa, in terms of emission: it is NBAPclay emission spectrum which shows a slight shift to higher energies in comparison to raw NBAP, while RhP-clay emission spectrum is absolutely the same as the raw RhP. Both NBAP-clay and RhP-clay show a well pronounced photo-activity. RhP-clay differs from NBAP-clay also because in its absorption spectrum there's a peak likely corresponding to the formation of J-aggregates (~548 nm), which are fluorescent according to the theory of exciton splitting and which are probably due to a regular organization of the intercalated dye induced by the presence of a regular array of alkylammonium molecules. The presence of a peak probably corresponding to the formation of non-fluorescent H-aggregates (~490 nm) is justified by the strong tendency of RhP molecules to aggregate – phenomenon which has already been reported in the literature [CAP04] and which clearly doesn't concern NBAP (unless at higher concentrations). A fluorescence absorption spectrum which indicates the presence of monomers and J-aggregates (likely dimers) into RhP-clay galleries is in perfect agreement with the results of molecular modelling recently obtained by Čapková et al. [CAP04].

### 6.6.4. Influence of Fluorescent Molecule Concentration

When dealing with fluorescence phenomena, fluorophore concentration is a key parameter because of the risk of quenching and the tendency to form supramolecular assemblies, which occasionally is a desired result but in any case has to be controlled, somehow. In the works of Esposito [ESP09], the influence of the concentration of fluorescent molecule on the results of clay photo-functionalization was studied. The cationic exchange processing in the presence of an amount of fluorescent molecule equivalent to 1MC – which turns out to surely be an excess of cationic organic dye was compared to a quarter of the former amount, i.e. 0.25 times the modifier concentration of C30B (0.25MC). In summary, clay photo-functionalization by cation exchange processing (D) with high concentrations of fluorescent molecule is possible (no significant fluorescence quenching seems to occur) but unnecessary, since most of the fluorophore excess is washed away and wasted. All the evidences collected so long indicate that cation exchange processing (D) with a smaller amount of an organic cationic dye like RhP (e.g. a relatively small percentage of the modifier concentration, *viz.* 0.25MC RhP) is sufficient to render the organoclay photo-active, even in the absence of a significant surfactant excess.

### 6.6.5. Organization of Organic Chains Inside the Clay Galleries

When organophilic clay already modified by alkylammonium ions are photofunctionalized by small fluorescent molecules like Rhodamine. Interlayer spacing could have increased in consequence of a kind of induced supramolecular organization but already exchanged molecules are the sole responsible for clay interlayer spacing or, because more likely, a rearrangement of the paraffinic configuration of alkylammonium molecules is induced by the presence of smaller (but still sterically-hindering) molecules on the absorbent surface. The presence of small molecules in addition to bigger molecules bearing long tallow fatty chains increases the characteristic paraffinic angle between the tallow chains and the silicate substrate from about 37° to 49° ca, as shown in Figure 11. A similar mechanism has been evocated to explain the crystalline swelling of organoclays in solution: intercalated organics are solvated by solvent molecules, which surround and support the aliphatic tails, causing a tilt with respect to the silicate surface [GAT04]. Molecular modeling confirmed that small molecules, when intercalated in clay galleries, lead to the tilting of bigger molecules, as shown by Čapková *et al.* [CAP04]. In the case of water molecules interacting with the xanthene planes of Rhodamine B (RhB) into MMT galleries, water molecules can be placed adjacent to the silicate layers and fill the empty spaces between RhB cations.

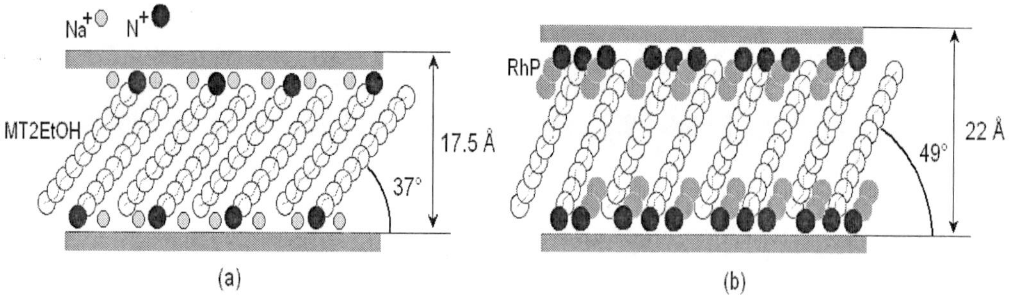


Figure 11. Possible explication for the increase of pristine interlayer spacing (a) following photo-functionalization with a cationic organic (b), under the hypothesis of a synergistic effect due to the presence of both small and bigger organic molecules cationic-exchanged within the silicate platelets [ESP08].

## 6.7. Conclusions

In this chapter, we have discussed the cationic exchange process which is the ordinary procedure to adsorb a chemical containing a charge-bearing group (i.e. an ionized molecule) on the internal surface of clay galleries. If the adsorbate is an organic molecule containing a charge-bearing group with a high affinity for clay immobilized anions (e.g. a quaternary ammonium salt), the cationic exchange process modifies clay nature (from hydrophilic to hydrophobic) to an extent which depends on the nature of the organic molecule. This treatment generates organically-modified clays, characterized by an improved compatibility with most of the polymers. If the adsorbate bears more than one function (e.g. cationic fluorescent dyes contain charge-bearing any fluorophore-bearing groups), cationic exchange process is a suitable method to render clays photo-active. Photofunctionalization can open large possibilities for all the modern on-line process monitoring techniques by conceiving a detection system capable of perform real-time monitoring of melt compounding of a polymer and a lamellar filler and sensitive to multiscale morphologies.

## Acknowledgements

The author wishes to acknowledge my former and present students as well as my colleagues with which I have had the pleasure to work. Many of their names appear in the references.

## References

[ALO06] Aloisi, GG; Costantino, U; Latterini, L. *J. Phys. Chem. Solids,* 2006, 67, 909-914.

[BAU01] Baumann, R; Ferrante, C; Deeg, FW. *J. Chem. Phys*., 2001, 114(3), 5781-5791.

[BOU02] Boucard, SJ; Duchet, J; Gerard, F; Prele, P; Gonzalez, S. *Macromol. Symp*, 2002, 194(1), 241-246.

[BOU03] Bourbigot, S; Vanderhart, DL; Gilman, JW. *J. Polym. Sci., Part B: Polym. Phys*, 2003, 41, 3188.
[BUJ06] Bujdak, J. *Appl. Clay. Sci*, 2006, 34, 58-73.
[BUR02] Burgentzlé, D; Duchet, J; Fillon, B; Jupin, A; Gérard, JF. *J. Colloid Interface Sci.*, 2004, 278, 26.
[BUR03] Burgentzlé. D. *Nouvelles formulations thermoplastiques ou réactives de revêtements nanocomposite à base de silicates lamellaires* Thèse matériaux: Institut National des Sciences Appliquées de Lyon, 2003 ; pp. 309.
[BUR04a]Bur, AJ; Roth, SC. *Polym. Eng. Sci.*, 2004, 44(5), 898-908.
[BUR04b]Bur, AJ; Roth, SC. *Polym. Eng. Sci.*, 2004, 44(4), 805-813.
[BUT01] Butzloff, P; D'Souza, NA; Golden, TD; Garrett, D. *Polym. Eng. Sci.*, 2001, 41, 10, 1794.
[CAL08] Calderon, JU; Lennox, B; Kamal MR. *Appl. Clay Sci.*, 2008, 40, 90.
[CAP04] Capkova, P; Mali, P; Pospisil. M. *J. Colloid Interface Sci.*, 2004, 277, 128-137.
[CAS03] Cassagnau, P; Melis, F; Bounor-Legare, V. *Polym. Eng. Sci,.* 2003, 43(4), 923-932.
[CAS05] Cassagnau, P; Courmont, M; Melis, F. *Polym. Eng. Sci.* 2005, 45, 926-934.
[CHI01] Chin, IJ; Thurn-Albrecht, T; Kim, HC; Russel, TP; Wang, J. *Polymer,* 2001, 42, 5947.
[COW58] Cowan, CT; White. D; *T. Faraday Soc.,* 1958, 54, 691-697.
[CZI04] Czimerova, A; Bujdak, J; Gaplovsky, A. *Colloids Surf. A, Physicochem. Eng. Aspects,* 2004, 243, 89-96.
[ENDO86] Endo, T; Sato, T; Shimada, M. *J. Phys. Chem. Solids,* 1986, 47(8), 799-804.
[ESP08] Esposito, A. *Visualisation de l'écoulement dans un système vis/fourreau. Suivi en ligne du mélange polymère/nanocharges* Thèse matériaux : Institut National des Sciences Appliquées de Lyon, 2008 ; pp. 270.
[ESP09] Esposito, A; Raccurt, F; Charmeau, JY; Duchet-Rumeau, J. In press in *J. Colloid Interface Sci.*, 2009.
[FUJ97] Fujita, T; Iyi, N; Kosugi, T. *Clays Clay Miner.,* 1997, 45(1), 77-84.
[GAR97] Garfinkel-Shweky, D; Yariv, S. *Clay Miner.,* 1997, 32, 653-663.
[GAT04] Gates, WP. *Appl. Clay Sci.*, 2004, 27, 1-12.
[GIE39] Gieseking, JE. *Soil Sci.,*1939, 47, 1-13.
[GIL04] Gilman, JW; Maupin, PH; Harris, RH. *Polym. Mater. Sci. Eng.*, 2004, 90, 717-718.
[HAC98] Hackett, E; Manias, E; Giannelis, EP. *J. Chem. Phys.* 1998, 108, 17, 7410-7415.
[HEN41] Hendricks, SB. *J. Phys. Chem.,* 1941, 45, 65-81.
[HER07] Hernandez, M; Sixou, B; Duchet-Rumeau, J; Sautereau, H. *Polymer*, 2007, 48, 4075.
[ING08] Ingram, SH; Petrick, RA; Liggat, JJ. *Polym. Inter.*, 2008, 57, 1206.
[IWA00] Iwasaki, M; Kita, M; Ito, K. *Clays Clay Miner.,* 2000, 48(3), 392-399.
[JOR49] Jordan, JW. *Mineral. Mag. J. M. Soc.*, 1949, 28, 598-605.
[KAN04] Kaneko, Y; Iyi, N; Bujdak, J. *J. Colloid Interface Sci.*, 2004, 269, 22-25.
[KIM06] Kim, NH; Malhotra, SV; Xanthos, M. *Micropor. Mesopor. Mater.,* 2006, 96, 29.
[KOR01] .Kornmann, X. *Synthesis and characterization of thermoset-layered silicate nanocomposites.* Doctoral thesis, Lulea University of Technology 2001; pp. 161.
[LAG86] Lagaly, G. *Solid State Ionics,* 1986, 22, 43.

[LAN94] Lan, T; Pinnavaia, TJ. *Chem. Mater*. 1994, 6, 2216.
[LAN95] Lan, T; Karivatna, PD; Pinnavaia, TJ. *J. Phys. Chem. Mater*., 1995, 7, 2144-2150.
[LEB99] LeBaron, PC;Wang, Z; Pinnavaia, TJ. *Appl. Clay Sci.* 1999, 15, 11-29.
[LEP02] Le Pluart, L; Duchet, J; Sautereau, H; Gerard, J. F. *J. Adhes*, 2002, 78, 645.
[LEP03] Le Pluart, L; Duchet, J; Sautereau, H; Gerard, J. F. *Macromol. Symp*, 2003, 194, 155-160.
[LEP05] Le Pluart, L; Duchet, J; Sautereau, H. *Polymer*, 2005, 46, 12267.
[LOP07] Lopez, Arbeloa, F; Tapia Estevez, MJ; Lopez Arbeloa, T. *J. Photochem. Photobiol. C : Photochem. Rev.*, 2007, 8, 85-108.
[LUC99] Luckham, PF; Rossi. S. *Adv. Colloid Interface Sci.*, 1999, 82, 43-92.
[MAC59] Mac Atee, JL. *Am. Miner.*, 1959, 44, 1230-1236.
[MAN91] Mani, S; Malone, MF; Winter, HH. *Macromolecules,* 1991, 24, 5451-5458.
[MAU04] Maupin, PH; Gilman, JW; Harris, RH. *Macromol. Rapid Commun*, 2004, 25, 788-792.
[MIT07] Mittal, V. *Euro. Polym. J.*, 2007, 43, 3727–3736.
[MIY00] Miyamoto, N; Kawai, R; Kuroda, K. *Appl. Clay Sci.*, 2000, 16, 161-170.
[NGO09] Ngo, TTV; Duchet-Rumeau, J; Whittaker, A; Gérard, JF. submitted in Polymer, 2009.
[OGA03] Ogawa, M; Ishii, T; Miyamoto, N. *Appl. Clay. Sci*, 2003, 22, 179-185.
[OGA97] Ogawa, M; Kuroda, K. *Bull Chem. Soc. Jpn.*, 1997, 70, 2593-2618.
[PAR03] Park, JH; Jana, SC. *Polymer,* 2003, 36, 8391.
[PAT93] Patzko, A; Dekany, I. *Colloids Surf., A* 1993, 71, 299-307.
[PIN00] Pinnavaia, TJ; Beall, G. W. *Polymer-clay nanocomposites* ; John Wiley & Sons, 2000, pp. 349.
[POS01] Pospisil, M; Capkova, P; Weiss, Z; Malac, Z; Simonik. J. *J. Colloid Interface Sci.*, 2002, 245, 126-132.
[ROW63] Rowland, RA; Weiss, EJ. *Clays Clay Miner.*, 1963, 10, 460-468.
[RUI04] Ruiz-Hitzky, E; Aranda, P; Serratosa, JM. In *Handbook of layered materials*; Taylor and Francis Group: NY, 2004.
[RYT00] Rytwo, G; Nir, S; Crespin, M. *J. Colloid Interface Sci.*, 2000, 222, 12-19.
[SAL08] Salleres, S; Lopez Arbeloa, F; Martinez, TV. *J. Colloid Interface Sci.*, 2008, 321, 212-219.
[SAS05] Sasai, R; Itoch, T; Iyi, N. *Chem. Lett.*, 2005, 34(11), 1490-1491.
[USU93] Usuki, A; Kawasumi, M; Kojima, Y; Okada, A; Kurauchi T; Kamigaito, O. *J. Mater. Res.*, 1993, 8, 1174-1178.
[VER05] Vermogen, A; Masenelli-Varlot, K; Seguela, R; Duchet-Rumeau, J; Boucard, S; Prele, P. *Macromolecules*, 2005, 38, 9661.
[VER07] Vermogen, A; Masenelli-Varlot, K; Vigier, G; Thollet, G; Duchet-Rumeau, J. *J. Nanosci. Nanotechnol.*, 2007, 1.
[VIE98] Vietze, U; Kraub, O; Laeri, F. *Phys. Rev. Lett.* 1998, 81(21), 4628-4631.
[WAN94] Wang, MS; Pinnavaia, T. *J. Chem. Mater*., 1994, 6, 468.
[XIE02] Xie, W; Xie, R; Pan, W; Hunter, D; Koene, B; Tan, L; Vaia, R. *Chem. Mater.*, 2002, 14, 4837.
[ZIL99] Zilg, C; Mülhaupt, R; Finter, J. *Macromol. Chem. Phys.*, 1999, 200, 661.

In: Polymer Nanocomposites: Advances in Filler Surface
Editor: Vikas Mittal
ISBN: 978-1-60876-125-8


*Chapter VII*

# Physical Adsorption on Clay Surface

***JiaQian Jiang**** [*]
Division of Civil, Chemical and Environmental Engineering, C5, Faculty of Engineering and Physical Sciences, University of Surrey, Guildford, Surrey, UK

## Abstract

This article reviews the essential feature of clay minerals, the possibilities of the clay modification and their use as potential adsorbents for the environmental remediation. The structure of clays is based on extensive sheets of silicon bonded with oxygen combined with metal oxide flat sheets (usually Al or Mg). Polar molecules are able to force the silicate layers apart because of relatively weak bonding. The swelling of clay minerals is an example of its electrical double layer properties. Properties of crystalline bonding and symmetry affect the charge on the crystal and affect the layer ratio of cations to anions. Cation exchange capacity is one of the most important characteristics of clay particles. Clays can be modified by metals, surface active agents (surfactants) or a mixture of the two. Montmorillonite and their modified products were assessed for the adsorption performance in the treatment of metals (e.g. Cu) and organic matter (e.g., phenol) in water. The results are discussed in the associated sections with the surface properties and structures of various clays.

**Keywords**: adsorption; clays; clay modification; montmorillonite; surface properties of clays; environmental remediation.

## 7.1. Introduction

Clay is one of the most common earth's minerals, which are the residue of weathering or hypothermal action. The classification and origin of clay depends on particle size, physical

[*] Corresponding author: j.jiang@surrey.ac.uk

characteristics, chemical compositions and common crystal structural characteristics. Clay'ssize is less than 2 micrometers with plastic properties when moist. Overall, clays are naturally hydrophilic minerals with layers of negatively charged platelets held tightly together by cations, which share the platelets to hold the stacks together. Hydrophilic material is attracted by strong dipole movements and is immiscible with organophilic material. Platelets are sheet-like structures whose dimensions are in two directions and are greater than the thickness of the particle.

Residual clays have been transported by natural processes and are found geologically positioned side-by-side with the igneous rocks from which they were formed. Sedimentary clays have been removed from their origin by natural processes. Within the United Kingdom, the china clays of Cornwall are considered the most important, termed "kaolin." The majority of these clays are found in Cornwall.

Overall, the two major groups that clay minerals fall into are 1) kaolinites and 2) montmorillonites. These clays are believed to come from feldspars of the granite group. When granite (igneous rock) is broken down, it first forms mica or illite that subsequently decomposes to kaolinites or montmorillonite shown below:

Feldspar → mica/illite → kaolinite
Feldspar → mica/illite → montmorillonite

The characteristics of clay minerals depend on the climate conditions prevailing at that point in geological history. For example, warm, humid conditions produced kaolinites while cooler, drier conditions produced montmorillonite. In the British Isles, kaolinites-type clays are predominant which show changes in climate to the current cold and wet conditions in the UK.

## 7.2. Physical Properties of Clays

Clay can take many forms and are composed of many different minerals. Clay-like properties include becoming sticky when mixed with water, mouldable when wet, while hard and brittle but retaining a shape when dry. If heated it becomes harder and unsusceptible to water. The swelling of clays minerals is an example of its electrical double layer properties.

Pore sizes of clays range in three different scales: sub 10 nanometre sized pores between platelet stacks; 10-100 nanometer sized pres between small platelet sub units; and micrometer sized pores with "free" water in between sub unit aggregates.

The general structure of a silicate mineral can be seen in Figure 1. It is structured in layers by the combination of sheets of articulated tetrahedral and octahedral elements. The basic layers for an octahedral 2:1 type of layer structure is generated by *Layer 1* in which two-thirds of the hydroxyls in one plane of the sheet are replaced by apical oxygen atoms of the tetrahedral sheet. Now the remaining OH ions of this sheet are in the centre of the hexagons formed by oxygen atoms. This generates a 1:1 unit structure defined by one octahedral and one tetrahedral sheet. In *Layer 2*, a second tetrahedral layer is added. The hydroxyls are replaced in a similar manner on the opposite side of the octahedral cation. This generates two tetrahedral sheets sandwiching a central octahedral sheet.

This specific arrangement of layers (as shown above) is found in the pyrophyllite-talc group, the smectites, the mica group, and the vermiculites. Within the variations of these groups, there are slight differences in the dimensions of the octahedral and tetrahedral sheets. When these sheets are stacked together, there may be internal stress. Substituting ions or rotating the tetrahedral can relieve the stress.

## 7. 2.1. Crystalline Properties

The majority of solids are crystalline where there is an array of atoms called a lattice. Each lattice is regular with a definite pattern of atoms repeated throughout the structure. A unit cell is the smallest group of atoms represented by a repeating pattern. The smallest entity in a lattice compound is a unit cell and should not be called molecule. Figure 2 shows a three-dimensional lattice as a simple geometrical Figure where the size of the unit cell is defined by the lengths of the three sides, *a, b*, and *c* as well as three angles α, β, and γ.

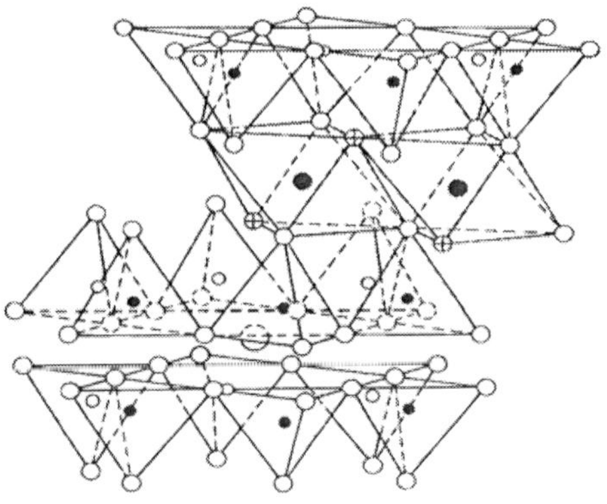

○ Oxygens ● Aluminum ⊕ Hydroxyls
• Silicons replaced by aluminums

Figure 1. Diagrammatic illustration of 2:1 layer lattice.

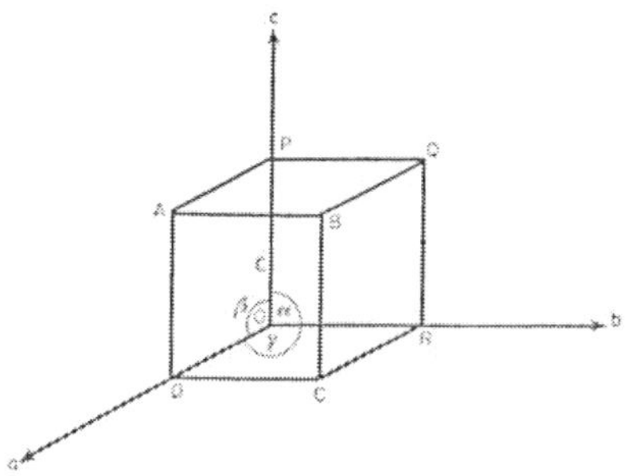

Figure 2. Crystallographic axes.

From Figure 2, it is observed that there are seven main crystal systems (Bravais space lattices). These systems can be further subdivided creating a total of fourteen possible cell unit types. The cube is the most symmetrical type of unit where $a = b = c$ and $\alpha = \beta = \gamma = 90^\circ$. The least symmetrical is the triclinic where no sides or angles are congruent. The shape of crystalline structures can become more complicated than the shape of the unit cell. Each face can consist of many planes of atoms that do not match the shape of the unit cell face.

It should be further noted that the method of classification of crystals is according to their symmetry which is defined as any convenient axis about which the crystal can be rotated. An object is symmetrical if on splitting, there are two identical sides. The more the object can be divided with identical sides, the more symmetrical the object is. A crystal with high symmetry can be divided along several axes through 360°. When rotated, the symmetry can be defined into thirty-two possible classes (e.g. fourfold, threefold, mirror plane, centre of symmetry and point groups symmetry).

In addition to the outer symmetry, the internal symmetry must be taken into account. This reveals that different types of atoms that can further account for a total number of symmetry operations (space groups) is 230. Therefore, the various planes of a crystal must be referred to in terms of axes to coincide the faces or the axes of symmetry. An axis can be chosen through various planes. The ratio of planes that intercepts an axis (and not the length) is the important concept. The Law of Rational Indices states "that the intercepts made by any other crystal plane on the three axes will bear a simple ratio of the chosen plane". The reciprocal of the intercepts is known as the Miller Indices which further shows the complicated levels of unit cells which can be found.

### 7.2.2. Crystalline Bonding

Crystalline bonds could be any following bonds or their combinations: electro-covalent, electro-negativity, polarity, hydrogen, and Van der Waals force; the type of bonds and the crystal co-ordination number give the physical shape of clays. The coordination number affects the building up of a crystal lattice. Within a crystal lattice, there is an ionic character on the atoms rendering their nature as ions. These ions have an ionic radius from the charged spheres. In clays, this is very important as oxygen is the most electro-negatively charged element occurring as $O^{2-}$. This gives a crystal lattice a form where the $Si^{+4}$ ion is surrounded by four oxygen ions (Figure 3). Within this sharing with silicon, one negative valence share from oxygen is free to link to other positive ions (such as $Al^{3+}$).

In order for the whole charge on the crystal to be neutral, the number of positive valence shares associated with each ion must equal the negative valence shares. The co-ordination number is the number of ions surrounding a given ion and is associated with the assumption that the ions are as close as they can get and are touching one another. The geometrical arrangement of the ions means that this is the maximum co-ordination number that can be in contact with the silicon. This number is expressed as a ratio of the radius of silicon ($Si^{4+}$ in the compound $SiO_2$) 0.39 Å to the radius of each oxygen ($O^{2-}$ surrounding the $Si^{4+}$) ion (1.40 Å). In other words, this is the ratio of the number of atoms closest to any given atom in a crystal.

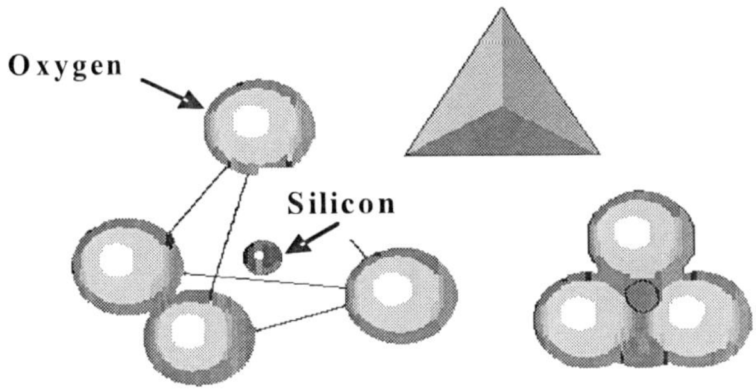


Figure 3. Silica tetrahedral $Si^{+4}$ building blocks of crystal lattice: geometric position of ions, lower view of base and triangular side-view of shape.

## 7.3. Surface Charge Properties

Surfaces charges can be developed onto a clay surface by adsorption of an ion. The solid acts as an electrode on which the ion is adsorbed. For example, on the surface of clays, $H^+$ and $OH^-$ can be adsorbed. In the case of a clay-aqueous system, determining the potential of the surface is a measure of the activity of ions which react with the mineral surface. When clay (adsorbent) adsorbs the adsorbent onto the surface, each layer attracts a specific ion. In Figure 4, the interaction of cations and anions with 1:1 kaolinite clay and a 2:1 smectite clay is shown.

### 7.3.1. Zero Point of Charge

The zero point of charge is the total charge from the cations and anions that is equal to zero (ZPC). At the ZPC, there is simultaneous adsorption of protons and hydroxyls and other potentially determining anions and cations. The total charge on the surface should be zero, while the number of opposing ions does not have to be equal. $H^+$ and $OH^-$ and complex ions formed from the bonding of $H^+$ and $OH^-$ are the potential determining ions in clays.

#### *Attached surface charge*

The kaolinite group has characteristic basal oxygen surface and will be used to illustrate the attachment of the surface charge. In this group, $O^{2-}$ ions bond with $H^+$ ions to form hydroxyls. The surfaces then can continue to react as either acids or bases with other protons or hydroxyls. Thus, the surface charge is pH dependent. The result is broken Si-O and Al-OH bonds along the surface of the clay crystals causing hydrolysis. Figure 4 shows the effects of a high pH and a low pH bonding on the surface.

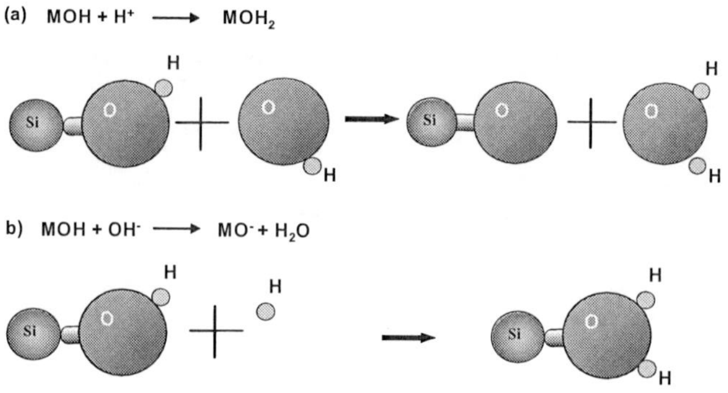


Figure 4. Bonding at clay surface at: (a) high pH (b) low pH.

Determining the potential at the surface is based on the movement of ions and can be expressed in the Nernst equation (7.1). In this equation, $E$ (electric potential) is related to the standard state electric potential ($E^o$) as follows:

$$E = E^o - \left(\frac{RT}{nf}\right) \ln\left(\frac{a}{a^o}\right) \tag{7.1}$$

where $E$ = electric potential; $E^o$ = standard state electric potential; $f$= Faraday constant; $n$ = number of molar equivalents of electrons transferred.

## *Isoelectric point*

At each zero point of charge, there is a corresponding pH called the isoelectric point, $pH_{ZPC}$. If the pH is below the $pH_{ZPC}$, then the solid has anion exchange capacity. At a low pH, the solution is in contact with the tetrahedral sheet basal oxygen surface containing excess protons. At the surface, the $SiOH^{2+}$ species dominates the surface of the clay and the anions in solution are attracted. The surface then exchanges anions limited to the anion exchange capacity. If the pH = $pH_{ZPC}$, the ions on the clay surface are balanced in charge of protons and hydroxyls, then there is no exchange capacity. At a pH above the $pH_{ZPC}$, the clay would have an excess of hydroxyls in contact with the solution at the basal oxygen surface of the tetrahedral sheet. The $OH^-$ species will dominate the surface causing the cations in solution attracted. Typical charges for clay minerals related to pH are summarised in Table 1.

**Table 1. Typical values of $pH_{ZPC}$ for clay minerals.**

| Mineral | $pH_{ZPC}$ |
|---|---|
| Gibbsite | 10 |
| Hematite | 4.2-6.9 |
| Goethite | 5.9-6.7 |
| Na-feldspar | 6.8 |
| Kaolinite | 2 - 4.6 |
| Montmorillonite | <2 - 3 |
| Quartz | 1 - 3 |

### 7.3.2. Electrical Double Layer

By adsorbing potential-determining ions, an electric double layer is developed. In kaolinite, there is an isomorphous replacement that gives a negative charge ($Si^{4+}$ replaced by $Al^{3+}$). Specific ion adsorption occurs when surfactant ions are specifically adsorbed. When clay is compacted, there are higher concentrations of cations and anions in the double layer with regards to the equilibrium solution.

Two layers are formed from the double layer. The inner layer is fixed and contains potential determining ions. The outer layer is a mobile diffuse layer of freely moving counter ions. Two representations of the ion distributions within the electrical double layer are the Gouy and Stern models.

### 7.3.3. Cationic Exchange Reactions

When clays are placed into a solution with a give electrolyte, the ions on the clay surface exchange with the ions of the electrolyte as described by equation 7.2:

$$\text{X-clay} + Y^{+} \leftrightarrow \text{Y-clay} + X^{+} \qquad (7.2)$$

This reaction is balanced proceeding from left to right depending on the nature of the ions X and Y, the nature of the clay, their relative concentrations and many secondary reactions. Some cations are more readily exchangeable than other cations and can be ordered by the lyotropic or Hofmeister series (Table 2). The anions similarly have a series that is based on the order by decreasing hydration.

#### *Cation exchange capacity (CEC)*

CEC is the maximum amount of cations that can be taken up by clay. The exchange reactions are usually isomorphous substitutions of $Al^{3+}$ for $Si^{4+}$ in the tetrahedral sheet and $Mg^{2+}$ and $Zn^{2+}$ for $Al^{3+}$ in the octahedral sheets. Basically, the CEC is determined by the leaching of clay with a chosen electrolyte such as ammonium acetate or an exchange resin. The electrolyte replaces the existing cation and owing to exchange of these cations, there is a negative surface charge on the surface of the clay often considered the attached charge. The charge imbalance is compensated by the exchangeable cations (like $Na^{+}$ or $Ca^{2+}$) at the surface of the clay. The total amount of these cations is defined as the cation exchange capacity and is expressed in terms of milli-equivalents per 100 g of dry clay.

**Table 2. Hofmeister series of adsorbed cations. (after Worral, 1986)**

| Hofmeister series: order of adsorbed cations | | | | | | | | | |
|---|---|---|---|---|---|---|---|---|---|
| Adsorbed more readily | | | | | Adsorbed less readily | | | | |
| H | Al | Ba | Sr | Ca | Mg | $NH_4$ | K | Na | Li |

Moreover, the CEC is enhanced by the properties of the clay that have been previously mentioned. The layer structure of the clay allows for the layers to expand when in contact with water exposing additional surface area of the mineral that is capable of cation adsorption.

Certain cations are adsorbed more strongly on clay surface over other cations. When Al or Fe are present as exchangeable ions, difficulties may be experienced in replacing cations and therefore the CEC will be low. In some specific clays like potassium montmorillonites, when the clay is dry, the fixation of cations is strong because the potassium fits tightly inside the structure of the sheets. In this case, the fixation is related closely to ionic size. In general, CEC increases with specific surface area as particle size decreases.

When using organic ions to exchange cations on the clay surface in aqueous solutions, amines may react with calcium clays and is expressed by:

$$R.NH_3^+ + \text{Ca-clay} \rightarrow R.NH_3\text{clay} + Ca^{2+}$$

where R is the alkyl or aryl group. Regardless of the nature of the original counter-ion, the clay maybe strongly flocculated by the adsorption of amine possibly because the amine is strongly attracted to the Stern layer.

In addition to cation exchange, anions can also be replaced in clays. This form of ion exchange in clays has been shown to reduce the alkalinity of NaOH and other bases, attributed to the adsorption of hydroxyl ions by the clay. Therefore, it is supposed that other anions can be adsorbed resulting in an exchange. The cause of this anionic adsorption is due to the direct chemical reactions with iron compounds in the soil rather than with the soil itself.

The rate of exchange of ions like $OH_3^+$ or $OH^-$ as counter-ions will depend on the pH. These ions exchange on sites that are either positive or negative (opposite $O_2^-$ ions or opposite $Al^{3+}$ or $Si^{4+}$ ions), the most frequent result is that the edges of clays are positive in their natural state. In other words, when no chemical dispersants have been added, the state of a clay surface is positive resulting in CEC as one of the most important characteristics of clay particles.

## 7.4. Modification of Montmorillonite

Montmorillonite, "fuller's earth" is an aluminium dioctahedral smectite clay mineral. It has a 2:1 configuration consisting of two silicon-oxygen tetrahedral sheets and one aluminium-oxygen-hydroxyl octahedral sheet. In the tetrahedral sheet, four oxygen atoms are coordinated with the silicon atom at the corner. The tetravalent silicon is sometimes replaced by trivalent aluminium. In the octahedral sheet, the aluminium atoms are coordinated with six oxygen atoms or hydroxyl groups which are located on the corners of the octahedron. The trivalent aluminium is replaced by magnesium, iron, zinc, nickel, lithium, or other cations. It has a high CEC compared to other clays.

Montmorillonite clays have excess negative charge generated in each triple sheet that is compensated by cations in the inter-triple sheet region. The capacity of montmorillonite to swell is an advantage. Dry montmorillonite adsorbs water between the triple sheets (interlayer region) about 4 layers of water molecules is adsorbed in each interlayer. As the clay swells, cation exchange can take place between the ions in solution and those in the interlayer when the triple sheets separate. In the interlayer, cations are $Na^+$ and $Li^+$ where interplate repulsion is greater and separation is easier.

Montmorillonite clays are commonly modified by polymeric aluminium or iron species, surfactant (e.g., HDTMA) or a mixture between the two (Jiang, et al, 2002, 2003, 2004). The modification is an isomorphous substitution of $Al^{3+}$ for $Si^{4+}$ in the tetrahedral sheet and $Mg^{2+}$ or $Zn^{2+}$ for $Al^{3+}$ in the octahedral sheet. This will result in a net negative surface charge on the clay. This negative charge imbalance is compensated by exchangeable cations (i.e. $Na^+$ or $Ca^{2+}$) at the clay surface. The total amount of these cations is called the cation exchange capacity.

Figure 5a shows a quaternary ammonium ion (e.g., HDTMA) with a long alkyl chain. This long chain makes it much more compatible with an organic matrix. Thus, organic cations are strongly adsorbed by silicate surfaces by ionic bonding. Long chains make the Van der Waals forces stronger and increase the adsorption of the matter in the matrix onto the modified clay surface. Figure 5b shows the montmorillonite structure repeats indefinitely extending to the left and the right. The green cations present on the clay are then replaced with another cation. Because the layered structure allows for expansion or swelling with water, more surface area is exposed which exposes more surface area capable of cation adsorption as shown in Figure 5c.

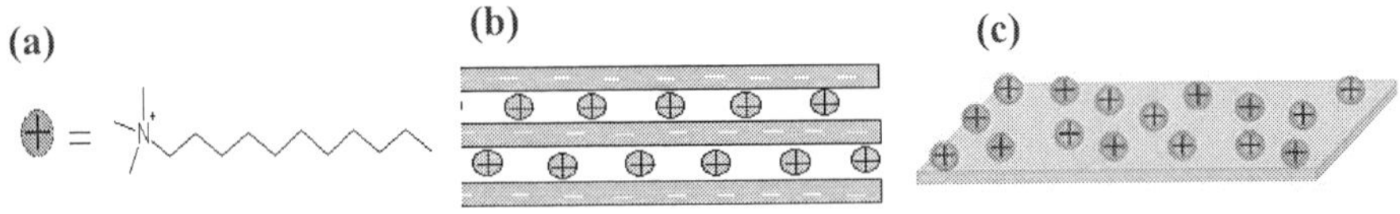


Figure 5. Modification of a montmorillonite; (a) quaternary ammonium ion with long alkyl chain, (b) montmorillonite structure and (c) cation exchange.

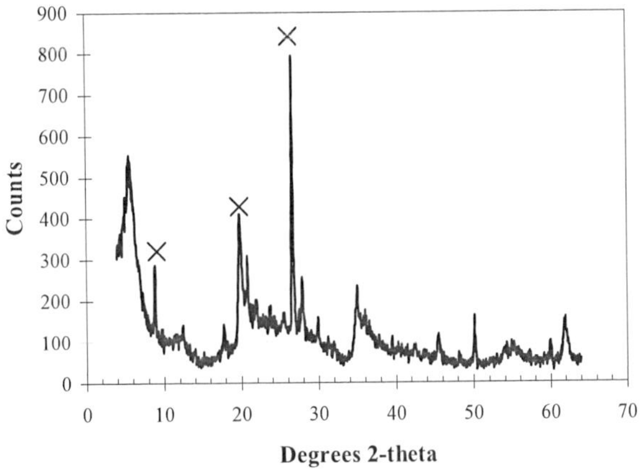


Figure 6. XRD trace of $Al_{13}$-polycation modified montmorillonite. Reprinted from Jiang, J. Q., Cooper, C. (2002), with permission from Elsevier.

Figure 6 shows an example of X-ray diffraction (XRD) traces for the modified montmorillonites. The peaks marked by (x) are the d001 reflections indicative of 2:1 swelling clays. The other peaks are impurities corresponding to quartz, plagioclase feldspar, illite and mica. Illite is a non-swelling 2:1 clay, mica is a non-swelling 2:1 phyllosilicate sheet silicate mineral and plagioclase is tectosilicate three-dimensional structure similar to zeolite framework. Table 3 summarizes the d001 reflection data for all of the samples analysed. From Table 3, it is clear that increase in basal spacing occurs in the presence of the metal polyoxocation/surfactant modified montmorillonites (except $Al_{13}$/HDTMA modified montmorillonite). This is due to two processes, first, the Fe and Al are known to form large aggregates with surfactants, larger than the polycationic species alone (Michot et al., 1993), therefore interaction between Fe/Al species with surfactants would result in larger *d*-spacing. Secondly, the polycationic species have probably undergone subsequent hydrolysis prior to XRD analysis, resulting in partial collapse of the interlayer spacing back to near the original value. However, polycations in the presence of surfactants are protected from hydrolysis resulting in the larger spacing being retained (Michot et al., 1993). $Al_{13}$/HDTMA modified montmorillonite (MMT) has a similar basal spacing (d001) to that of starting MMT, and the reason of this needs to be studied in future work.

**Table 3. Basal spacing for modified montmorillonites. Reprinted from Jiang, J. Q., Cooper, C. (2002), with permission from Elsevier.**

| Type of Montmorillonites (Mt.) | 2θ value | Basal spacing (d001)/Å |
|---|---|---|
| Starting (Mt.) | 5.7 | 15.5 |
| $Al_{13}$ -modified (Al_Mt) | 5.4 | 15.9 |
| Fe –modified (Fe_Mt) | 5.8 | 15.3 |
| $Al_{13}$/Fe–modified (Al_Fe_Mt) | 5.9 | 17.8 |
| HDTMA–modified (R_Mt) | 6.2 | 14.4 |
| Fe/HDTMA –modified (Fe_R_Mt) | 5.0 | 18.1 |
| $Al_{13}$/HDTMA –modified (Al_R_Mt) | 5.75 | 15.4 |
| Al/Fe-HDTMA –modified (Al_Fe_R_Mt) | 5.1 | 18.0 |

## 7.5 Adsorption of Cu and Phenol in Water Treatment

The adsorption behaviour, especially the adsorption capacity of the clays may be affected by the interlayer spacing and the surface affinity of the clays, whereas which factor would be the predominant is under discussion. Figure 7 – 8 show the adsorption isotherms of Cu onto the raw and modified montmorillonite K10 and KSF clays respectively, the adsorption coefficient (k) and the Freundlich constant (a) are calculated in Table 4. The importance of the surface properties was emphasized by the adsorption results of montmorillonite KSF and K10 clays. As shown in Table 4, the k value had a sequence of Fe/Al modified clays > Fe/Al-HDTMA modified clays > HDTMA modified clays > raw clays after pre-treatment > raw clays before pre-treatment. Such a result cannot be explained by structural effects because the k value sequence was almost on the contrary of the d-spacing sequence. It is apparent that the Fe/Al species can significantly improve the Cu adsorption performance due to that polymeric Fe/Al modified clays have a highest affinity for the metal ions. Hence the adsorption of Cu onto clays would be determined by both basal spacing and polymeric Fe/Al surface content of the clays; the effects of d-spacing as well as surface properties could play equal roles in adsorbing metal species.

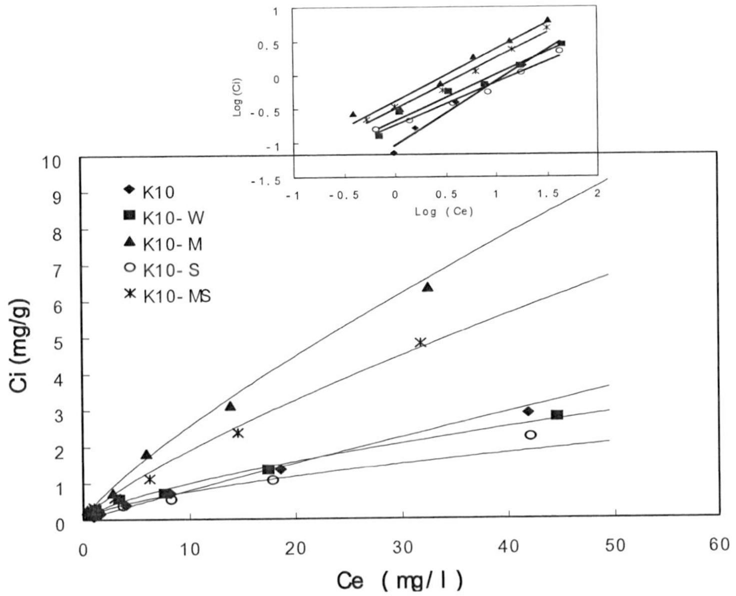


Figure 7. Adsorption isotherm of Cu onto the raw and modified montmorillonite K10 clays. Reprinted from Jiang, J. Q., Zeng, Z. (2003), with permission from Elsevier.

**Table 4. Freundlich isotherm constants for the adsorption of Cu onto modified clays. Reprinted from Jiang, J. Q., Zeng, Z. (2003), with permission from Elsevier.**

| Clay | Adsorption coefficient (k) | Freundlich constant (a) | Correlation coefficient (R) |
|---|---|---|---|
| K10 | 0.09 | 0.94 | 0.99 |
| K10-W | 0.21 | 0.68 | 0.99 |
| K10-M | 0.40 | 0.81 | 0.99 |
| K10-S | 0.18 | 0.63 | 0.99 |
| K10-MS | 0.31 | 0.79 | 0.98 |
| KSF | 0.43 | 0.84 | 0.99 |
| KSF-W | 0.50 | 0.46 | 0.99 |
| KSF-M | 1.11 | 0.62 | 0.99 |
| KSF-S | 0.47 | 0.62 | 0.99 |
| KSF-MS | 0.77 | 0.64 | 0.99 |

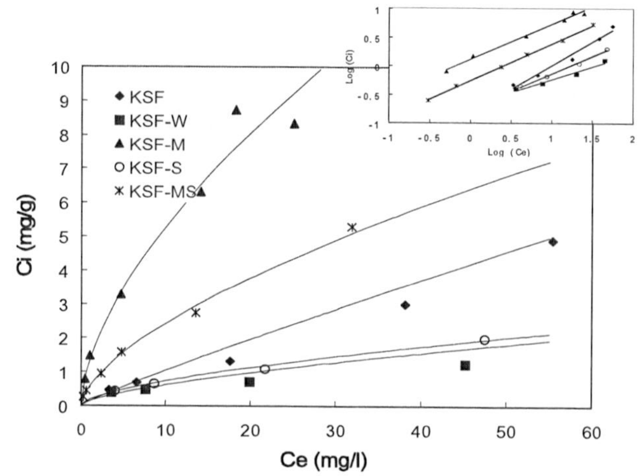


Figure 8. Adsorption isotherm of Cu onto the raw and modified montmorillonite KSF clays. Reprinted from Jiang, J. Q., Zeng, Z. (2003), with permission from Elsevier.

The phenol adsorption results indicated that raw clays and polymeric Fe/Al modified clays could hardly adsorb phenol, as shown in Figures 9 and 10. According to the adsorption coefficients summarized in Table 5, the k value of these clays were so small that they could be roughly regarded as zero, thus, any fluctuation in these k value after pre-treatment or modification with polymeric Fe/Al could be neglected. For montmorillonite K10 and KSF, it was in common that the HDTMA modified clays and Fe/Al-HDTMA modified clays were able to adsorb phenol while the adsorption capacity of the former was larger than that of the latter. This cannot be explained by basal spacing effects because in the cases of K10 and KSF, the d-spacing of HDTMA modified clay was even smaller than that of Fe/Al-HDTMA modified clay – suggesting the importance of the surfactant modifiers and the resulting surface properties.

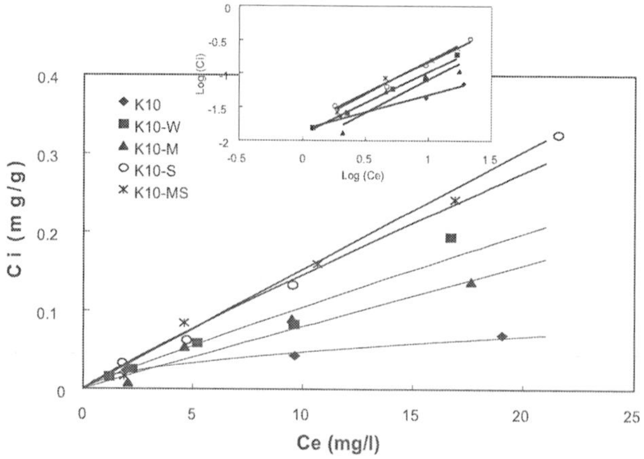


Figure 9. Adsorption isotherm of phenol onto the raw and modified montmorillonite K10 clays. Reprinted from Jiang, J. Q., Zeng, Z. (2003), with permission from Elsevier.

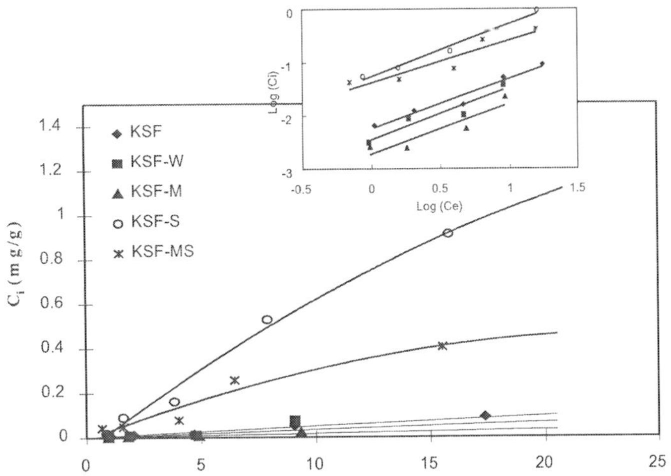


Figure 10. Adsorption isotherm of phenol onto the raw and modified montmorillonite KSF clays. Reprinted from Jiang, J. Q., Zeng, Z. (2003), with permission from Elsevier.

The possible reason of this could be that the phenol molecules are larger than Cu ions, and the basal spacing is too small to affect the phenol adsorption performance, therefore, the hydrophobic surface properties of the surfactant modified clays play the dominant role in the adsorption of phenols in water.

**Table 5. Freundlich isotherm constants for the adsorption of phenol onto modified clays. Reprinted from Jiang, J. Q., Zeng, Z. (2003), with permission from Elsevier.**

| Clay | Adsorption coefficient (k) | Freundlich constant (*a*) | Correlation coefficient (R) |
|---|---|---|---|
| K10 | $1.46\times10^{-2}$ | 0.51 | 0.99 |
| K10-W | $1.25\times10^{-2}$ | 0.92 | 0.99 |
| K10-M | $0.83\times10^{-2}$ | 0.98 | 0.95 |
| K10-S | $1.73\times10^{-2}$ | 0.92 | 0.99 |
| K10-MS | $1.54\times10^{-2}$ | 0.99 | 0.99 |
| KSF | $0.58\times10^{-2}$ | 0.93 | 0.98 |
| KSF-W | $0.36\times10^{-2}$ | 0.97 | 0.95 |
| KSF-M | $0.19\times10^{-2}$ | 0.93 | 0.92 |
| KSF-S | $5.34\times10^{-2}$ | 0.99 | 0.99 |
| KSF-MS | $4.19\times10^{-2}$ | 0.79 | 0.94 |

## 7.6. Conclusions

The structure of clays is based on extensive sheets of silicon bonded with oxygen combined with metal oxide (usually Al or Mg). Polar molecules are able to force the silica layers apart because of relatively weak bonding. The swelling of clay minerals is an example of its electrical double layer properties. Crystalline bonding and symmetry affect the charge on the crystal and affects the layer ratio of cations to anions. Cation exchange capacity is one of the most important characteristics of clay particles. Clays can be modified by metals, surface active agents (surfactants) or a mixture of the two.

Montmorillonites K10 and KSF were modified by the polymeric Fe/Al, surfactant (HDTMA), and Fe/Al-HDTMA species, respectively and were assessed for the adsorption performance in the treatment of metals (e.g., Cu) and organic matter (e.g., phenol) in water. The results indicated that both basal spacing and surface properties of clays would affect the water and wastewater treatment performance. In general, adsorption mechanisms of modified clays were dependent on the specific surface properties. For adsorbing inorganic/metal species (e.g., Cu), the specific chemical complexation capability of polymeric Al/Fe species plays a major role whilst for adsorbing organic contaminants (e.g., phenol), the hydrophobic property of the surfactant (HDTMA) attributed to the better adsorption performance.

## Acknowledgement

The work described here was partly contributed by the author's postdoc researchers and research students, Drs. C. Cooper; Z. Zeng, G. Kim and Ms. L. V. Vlaservich. The financial support for the various projects from the UK EPSRC is also highly appreciated.

## References

[1] Brindley, GW; Brown, G. *Crystal structures of clay minerals and their X ray identification;* Mineralogical Society: London, 1980, pp 495.

[2] In *The X-ray identification and crystal structures of clay minerals*; Brown, G; Ed; Jarrold and Sons Ltd.: Norwich, 1961.

[3] Cooper, C; Jiang, JQ; Ouki, S. *J. Chem. Technol. Biotechnol.,* 2002, 77, 546-551.

[4] In *Clay Mineralogy*; Grim, R. E; Ed; McGraw-Hill: New York, 1953.

[5] Jiang, JQ; Cooper, C. *Chemosphere,* 2002, 47, 711-716.

[6] Jiang, JQ; Zeng, Z. *Chemosphere,* 2003, 53(1), 53–62.

[7] Jiang, JQ, Cooper, C. *Environ. Eng. Sci.,* 2003, 20, 581-587.

[8] Jiang, JQ; Zeng, ZQ; Pearce, P. *Water, Air, Soil Pollution,* 2004, 158(1), 53–65.

[9] Jiang, JQ; Kim, CG. *Separation Sci. Technol.,* 2008, 43, 1677- 1686.

[10] Kim, CG; Jiang, JQ; Choi, CH. In *Trends in Water Pollution Research*; Livingstone, J. V; Ed; Nova Science Publishers: New York, 2005, pp 75-93.

[11] Worrall, WE. *Clays and Ceramic Raw Materials*; Elsevier Applied Science Publisher: London and New York, 1986.

[12] Zeng, Z; Jiang, JQ. *Int. J. Environ. Studies,* 2005, 62(4), 403–414.

In: Polymer Nanocomposites: Advances in Filler Surface
Editor: Vikas Mittal
ISBN: 978-1-60876-125-8


*Chapter VIII*

# Role of Clean Clay Surface on Composite Properties

*Vikas Mittal*[*]
Department of Chemistry and Applied Biosciences, Institute of Chemical and Bio Engineering, Swiss Federal Institute of Technology, ETH Hoenggerberg, 8093, Zurich, Switzerland

## Abstract

Polymer nanocomposites require the compatibilization of the organic and inorganic phases in order to achieve optimal microstructure of the composites and subsequently improved composite properties owing to the delamination of the filler. Therefore, the surface modification of the filler surface is achieved which makes it compatible with the polymer matrices. The nature of the achieved modified filler surface as well as its cleanliness has tremendous impact on the composite properties. The cleanliness of the filler surface is analyzed by using high resolution thermogravimetric analysis where the filler modification molecules not bound to the filler surface and present only as pseudo bilayer degrade at lower temperature and the absence of low temperature degradation peak in the thermogravimetric analysis thus would confirm the achievement of clean filler surface. The X-ray diffraction is not suitable to ascertain the cleanliness of the surface as the clean fillers as well as the fillers with excess modification have been observed to have similar basal plane spacing values. The removal of the excess modification is necessary as it may interact with the polymer matrix in more than one ways. At high temperature, its low temperature degradation can impact the thermal properties of the composites, it may also have unwanted chemical reaction with the polymer matrix thus degrading the polymer molecular weight, it may also gelate the dispersion of the filler with polymer in solvents thus destabilizing the system. The presence of excess amount of surface modification can also lead to increased free volume at the interface owing to the non-optimum development of the interface where the organic phase may not be fully compatible with the surface modification. The

[*]Correspondence author: BASF SE, Polymer Research, Ludwigshafen, Germany vikas.mittal@basf.com

commercially available modified montmorillonites have been conformed confirmed to contain an excess of surface modification molecules, therefore, their use as it is may have additional interactions in the composites. The presence of the excess of the surface modification molecules on the surface was found to have significant effect on the composite properties, both mechanical and permeation properties were observed to decrease as a result of the excess surface modification. The microstructure as studied by transmission electron microscope however did nit not show any difference in the systems with or no excess thus indicating that the developed microstructure may not be affected by the presence of these excess ions, but these ions surely change the interfacial interactions leading to deterioration in the composite properties.

**Keywords**: Surface modification, excess of filler modification, high resolution thermogravimetric analysis, thermal degradation, interfacial interactions, permeation, mechanical properties, microstructure.

## 8.1. Introduction

The surface modification step of the inorganic filler is the most important parameter to ensure the synthesis of nanocomposites with uniform and nanoscale distribution of the filler [1-6]. The generation of microstructure in the polymer nanocomposites is a function of the interactions of the surface modification with the polymer matrix. When the interactions between the organic and inorganic phases are not positive e.g. incompatibility between the polar polymer and non-polar surface modification, the generated interface is not optimum and subsequently the composite properties are also negatively impacted [7]. Similarly, when the interactions between the organic and inorganic phases are zero, there is no adhesion at the interface and the mechanical properties of the composites are affected accordingly. The thermal stability of the surface modification is also a factor which contributes to the interactions between the two phases. If the surface modification is not thermally stable, then it can lead to unwanted side reactions with the polymer matrix leading to reduction in molecular weight as well as deterioration of the interface between the phases. Thus the importance of surface modification in leading to the better composite properties is very significant. The cleanliness of the filler surface is also another factor which is related to the importance of surface modification. If the filler surface is not clean, i.e. if there is a presence of excess surface modification molecules on the filler surface, these can either chemically interact with the polymer matrix or can degrade at low temperatures owing to their low temperature stability thus inducing unwanted side reactions with the polymer matrix. The commercially available modified inorganic fillers have always occasionally an excess of the surface modification present on their surface, therefore it is quite possible than many of the reported effects were affected by this excess [8]. It may also be the reason for the observation of no particular trend in the properties of especially polyolefin nanocomposites. Therefore, oOne must be careful to confirm if there is a presence of such an excess on the filler surface. Generally the thermogravimetric analysis is used to attain information on such issues, but the thermogravimetric analysis also leads to the information on the excess of the surface modification only if the high resolution measurements are carried out. In the high-resolution measurements, the temperature of the sample is not raised until the weight loss at a particular temperature is completed [9]. This leads to accurate quantitative analysis of the presence of excess of the surface modification present on the surface of the filler. Figure 1 shows the representation of a montmorillonite surface with excess of surface modification molecules forming a pseudo bilayer which is present only as adsorbed in between the surface modification molecules and not bound ionically to the surface. The clean surface counterpart is also shown in the Figure where the excess has been either washed off or the exchange has been carried out in a way that no such pseudo bilayer is formed.

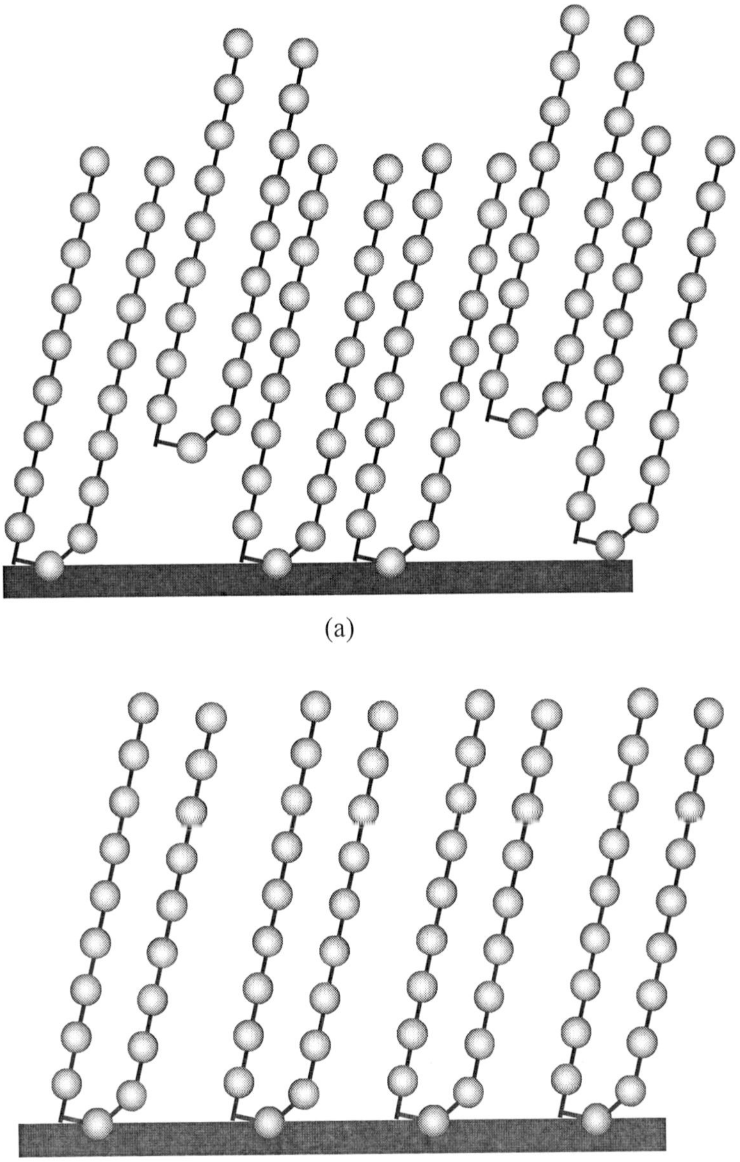

(a)

(b)

Figure 1. Representation of (a) montmorillonite surface with pseudo bilayer and (b) montmorillonite with clean surface.

## 8.2. Analysis of the Cleanliness of the Filler Surface with TGA and XRD

Thermogravimetric analysis is the most important tool to study the cleanliness of the filler surface. Therefore, it is absolutely important to carry out the thermal analysis of the filler after surface modification as well as after any other surface reaction in order to ascertain the presence or absence of free surface modification molecules or in general to know about the temperatures till which the filler modification would be thermally stable. Thermogravimetric analysis of the composites is also necessary to be performed in order to analyze the possible effects of partial thermal degradation of the filler surface modification during compounding at high temperature in on the thermal stability of the polymer matrix. As mentioned above that in order to ascertain the state of the filler surface, high resolution should be used in the thermogravimetric analysis owing to the fixing of temperature at one point until the weight loss at that point is not over. The conventional non-high resolution thermogravimetric analysis can sometimes lead to significant shifting of the peak positions as the temperature is not held constant during the event of weight loss at particular temperature. It has been experimentally observed that filler surface has a linear increase in the amount of organic matter exchanged on the surface and the amount of surface coverage with the organic modification when the amount of surface modification is increased in the system during the filler modification process. However, the amount of surface coverage does not increase linearly as a function of amount of surface modification after the 80% of the surface has been covered [10]. Further addition of surface modification thus only leads to the generation of a pseudo bilayer of the surface medication modification molecules which are only physically trapped in between the surface modification molecules which are ironically bound to the clay surface, whereas these pseudo bilayer molecules are not bound to the clay surface and have a negative impact on the thermal stability of the surface modified filler. This leads to the reduced properties of the composites owing to the negative impact of on the polymer matrix as well as non-optimum development of the interface between the organic and inorganic phases.

Not only that, the stability of the dispersions of the fillers with the polymer in solvent can be affected badly if there is a presence of excess surface modification molecules on the filler surface. In one of the studies, a phenomenon of ‘‘degellation’’ or ‘‘deexfoliation’’ of the organo-montmorillonite, containing bis(2-hydroxyethyl) methyltallowammonium as surface modification, was reported on addition of polar additives such as amines as crosslinking agents [11]. It was suggested in this study that bifunctional amine molecules may be able to bridge the silicate layers or the N–H groups in the primary and secondary amines may be polar enough to cause re-aggregation of silicate layers. The other possible reason for this behavior may be the presence of excess molecules of surface modifier on the clay surface, which are not be ionically attached to it. These molecules being free on the clay surface may easily interact with the amine or other polar additives thus coagulating the whole dispersion [10]. The dispersions were absolutely stable when the filler surface was self-modified in the laboratories and with the help of high resolution thermogravimetric analysis, it was made sure that it did not contain any excess of surface modification.

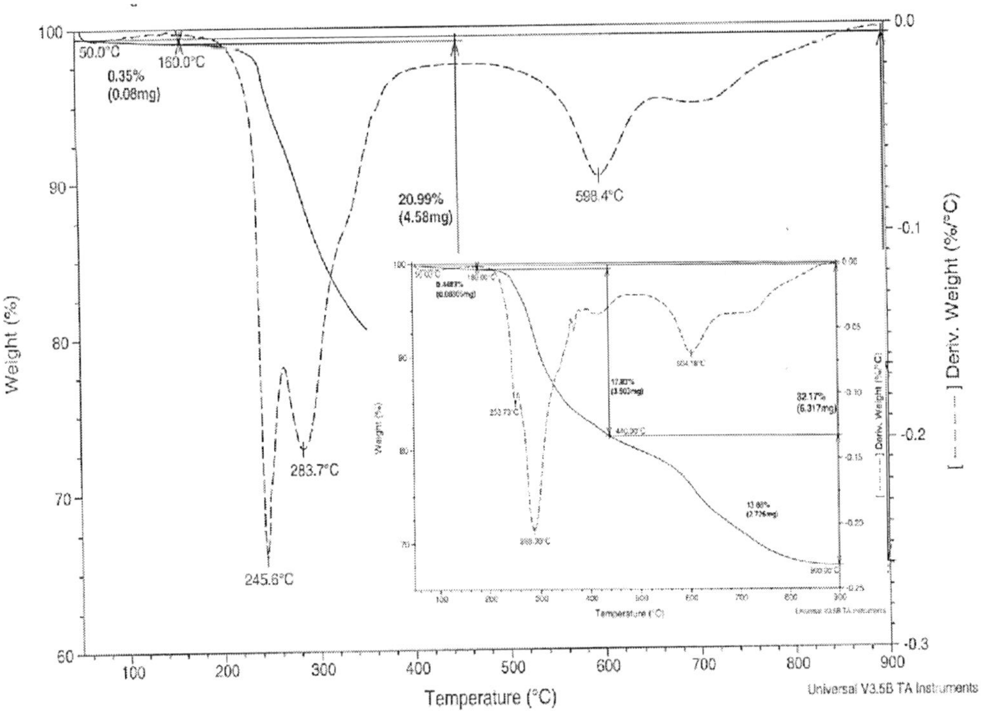


Figure 2a. TGA plots of the commercially available dioctadecyldimethylammonium modified clay and in inset is the lab modified montmorillonite (cation exchange capacity of 680 μeq.g$^{-1}$).

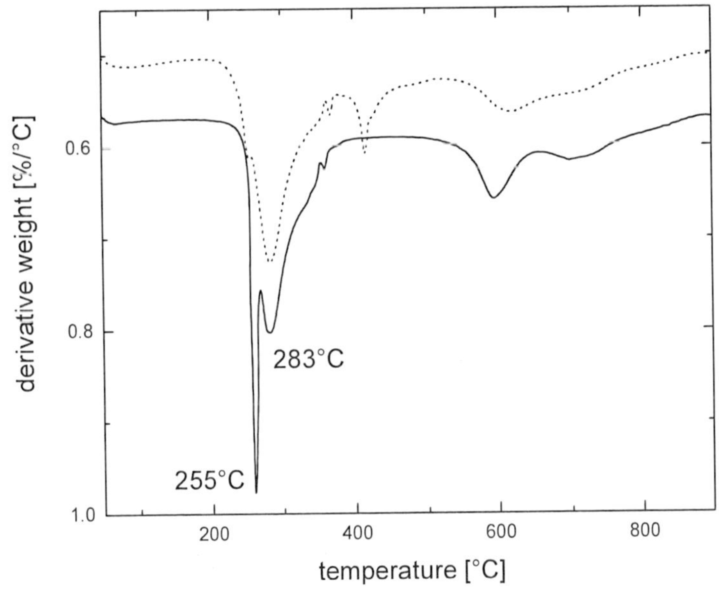


Figure 2b. TGA thermograms of the montmorillonites modified with dioctadecyldimethylammonium, commercial (dark line) and self modified (dotted line) montmorillonites (cation exchange capacity of 880 μeq.g$^{-1}$)..

As mentioned earlier that the commercially available surface modified montmorillonites have always occasionally been observed to contain an excess of surface modification molecules, therefore one should access the exact thermal state of the system by the use of high resolution theromgravimetric analysis. As an example, in Figure 2a are shown the TGA thermogrmas of the commercial montmorillonite modified with dioctadecyldimethylammonium modification exchanged on a montmorillonite with a cation exchange capacity of 680 μeq.g$^{-1}$. There is a sharp peak of thermal degradation at 246°C owing to the presence of excess surface modification on the filler surface. The modification molecules ironically bound to the filler surface have peak degradation temperature of 284°C. In the inset of this Figure is also shown the montmorillonite modified with the same modification in the lab. The cCare was taken in order not to add an excess of the surface modification molecules during the cation exchange reaction as well as the filler surface was cleaned a number of times after the surface modification in order to clean the surface off from any of the modification molecules present as pseudo bilayer. The TGA thermogram thus shows only one peak of thermal degradation at 288°C and the low temperature degradation related to the presence of excess surface modification could be almost eliminated. Similarly, Figure 2b also compares the lab modified and commercially procured montmorillonites also modified with dioctadecyldimethylam-monium modification, but on a montmorillonite with a cation exchange capacity of 880 μeq.g$^{-1}$. In this case, also in the lab modified montmorillonite the low temperature degradation peak of 255°C in the commercially modified montmorillonite was absent indicating the surface was optimally modified. The self-modified montmorillonites may require 4-5 cycles of cleaning in order to completely wash off the unattached surface modification molecules. In one such cycle, the clay is suspended in a solvent which swells the clay extensively. The clay is then rotated for some time at a temperature of generally 70°C and during this time, the clay is also subjected to sonication as well as shear mixing in order to disperse it uniformly in the solvent. Afterwards, the clays is filtered and is also extensively washed on the filter. The TGA of the cleaned clay is then carried out. Depending on the degree of cleanliness of the surface, the clay is ether used for the composite preparation or is subjected to another cleaning cycle. The amount of surface coverage in the clean clay should also be near to 100% as the complete organophilization of the surface is also equally necessary for the composite properties as the clean filler surface. Figure 3 compares the TGA thermograms during different cycles of cleaning. The clay was modified with dioctadecyldimethylammonium modification and the gradual cleaning of the surface after every cycle is clearly observed in the Figure. The importance of cleaning of the filler surface is also significant when there are some surface reactions on the surface. As an example, physical adsorption of the polymer chains are carried out on the surface modified filler in order to completely cover the filler surface so as to achieve its nanoscale dispersion in the polymer matrices. The polymer during this process can physically adsorb on the filler surface or a part of it can also entangle with the already immobilized ammonium ion chains on the surface, which would then have lower degradation temperature. As shown in Figure 4, the cleaning of the surface off the free polymer chains is also helpful in order to achieve optimal performance from the filler. One should also be careful that though the excess of the surface modification in the clay interlayers can be detected in TGA, but mostly X-ray diffractograms are silent to this phenomenon as shown in

Figure 5. No change in basal spacing of the clay is observed in the commercial and lab modified clay except from the unrelated change in intensity.

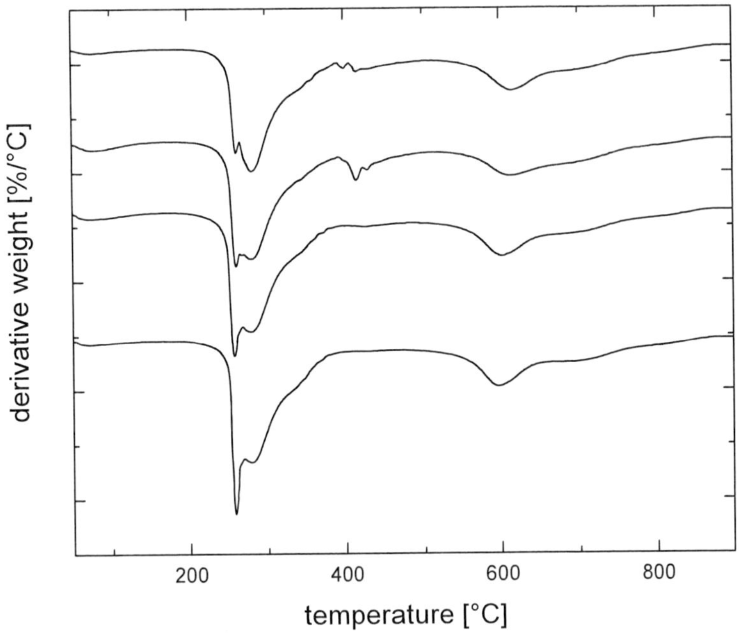


Figure 3. The TGA thermograms indicating the effect of cleaning of the filler surface. The uppermost curve represents the filler after 4-5 washing cycles.

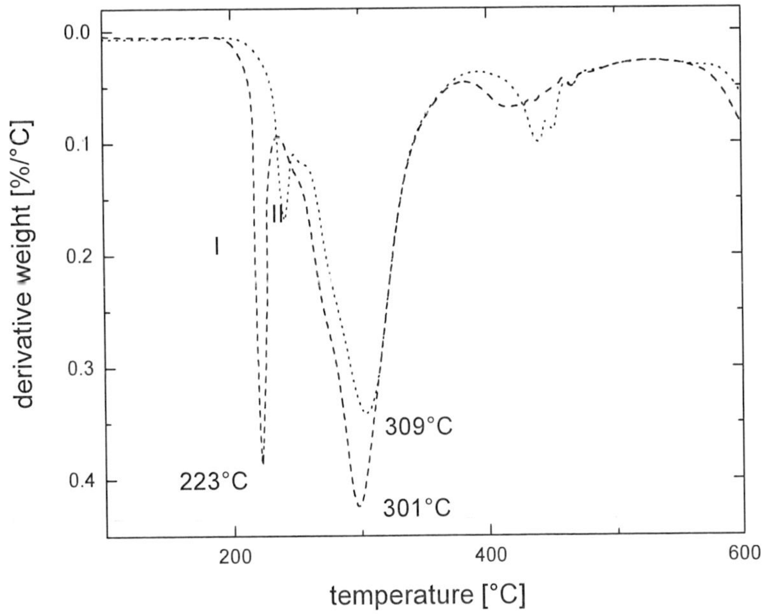


Figure 4. Role of cleaning during the physical absorption of the polymer chains on the organically modified montmorillonite, I: before cleaning and II: after cleaning.

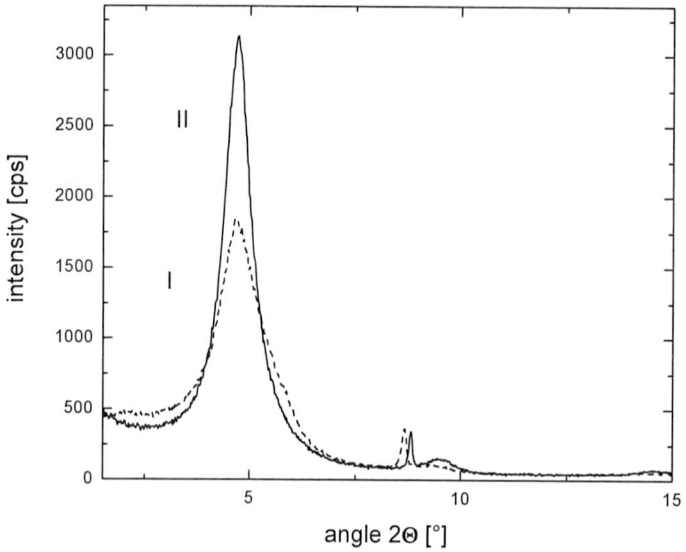


Figure 5. X-ray diffractograms of the surface modified montmorillonites, I: commercial and II: lab modified montmorillonite.

## 8.3. Impact on Composite Properties

Apart from decreasing the thermal stability of the surface modification and bringing in the related negative effects on the composites as well as destabilization of the suspensions of the polymer chains in solvents, the effect of presence of excess of the surface modification on the composite properties is also necessary to be quantified. Osman et al. studied the effect of excess amount of surface modification used to modify the surface of calcium carbonate particles [12]. The composites of such particles with low density polyethylene were observed to have decrease in the mechanical properties except stiffness. It was suggested that optimum amount of filler surface modification should be used in the modification process and excess should be avoided. Oxygen permeation properties of epoxy nanocomposites with montmorillonites having an excess of surface modification were compared with the properties of the nanocomposites with montmorillonites having no excess surface modification [8]. A very significant effect of the presence of the excess was observed on the permeation properties which are very sensitive to the interface development and minor extents of interface deterioration leads to a significant negative impact of on these properties. The properties of the composites with commercial and the lab modified montmorillonites are compared with each other in Figure 6. The inorganic volume content was fixed at 3.5 vol% in order to make the composite properties comparable and all the synthesis processes were always the same.

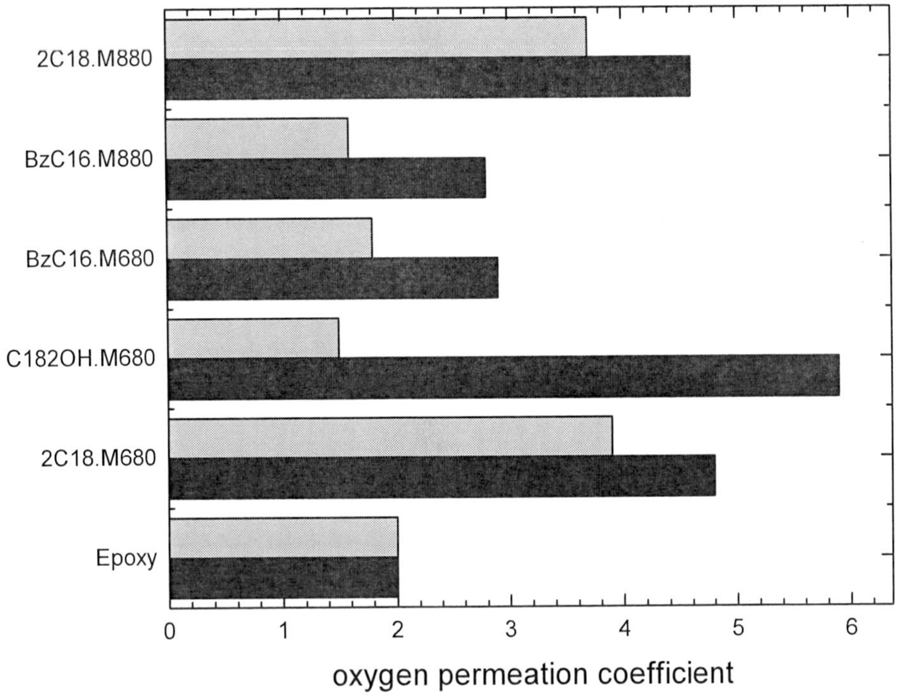


Figure 6. Comparison of the oxygen permeation properties of the epoxy nanocomposites with commercially modified montmorillonite (black bar) and lab modified montmorillonite (grey bar).

The montmorillonites with two different cation exchange capacities were used in order to quantify the effect of cation exchange capacity. The epoxy nanocomposite without filler had value of oxygen permeation of 2 $cm^3.\mu m/m^2.d.mmHg$. In the case of dioctadecyldimethylammonium modified montmorillonites (2C18.M680 and 2C18.M880), the composites with commercial or lab modified montmorillonites had high oxygen permeation values as compared to the pure epoxy polymer though the latter had lesser increase in permeation with respect to the pure polymer. The increase in the permeation is a result of the incompatibility of the non-polar surface modification with the polar polymer matrix which leads to an increase of the free volume at the interface which subsequently causes increased oxygen permeation. The composites with benzylhexadecyldimethylammonium modified montmorillonites (BzC16.M680 and BzC16.M880), the effect of clean filler surface was obvious as the presence of cleaner filler did not deteriorate the oxygen permeation and a decrease in the oxygen permeation was observed, which was higher in the case of the composite with higher cation exchange capacity clay. A further significant improvement in the case of clean filler composites was observed in the case of C182OH modified montmorillonites. Thus, it is clear that the clean surface of the filler is very important requirement of for the achievement of optimum composite properties. The TEM micrographs of the composites with commercial and self modified montmorillonites were practically indistinguishable and in both the cases, a mixed morphology was observed where a mixture of intercalated and exfoliated filler platelets was observed. It further indicated that the microstructure development may not be affected by the presence of these free ions on the surface, but their interactions with the polymer matrix leads to different interactions at the interface which causes increase in the oxygen permeation owing to the increased free volume at the interface.

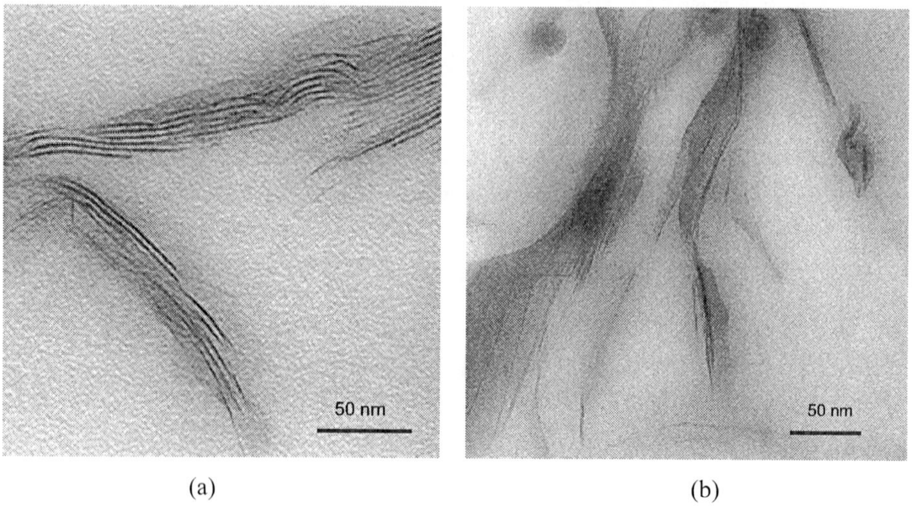


Figure 7. TEM micrographs of the 3.5 vol% composites with (a) commercially modified and (b) self modified montmorillonites.

# References

[1] Theng, BK; G. *The Chemistry of Clay-Organic Reactions;* Adam Hilger: London, 1974.

[2] Lagaly, G. *Developments in Ionic Polymers;* AD. Wilson, HJ. Prosser, (Eds.); Elsevier Applied Science Publishers, London/New York, 1986.

[3] Pinnavaia, TJ. *Science*, 1983, 220, 365.

[4] Giannelis, EP. *Adv. Mater*, 1996, 8, 29.

[5] Lagaly, G. *Solid State Ionics*, 1986, 22, 43.

[6] Lagaly, G; Benecke, K. *Colloid Polym. Sci*, 991, 269, 1198.

[7] Osman, MA; Mittal, V; Morbidelli, M; Suter, UW. *Macromolecules*, 2003, 36, 9851.

[8] Mittal, V. *J. Mater. Sci.*, 2008, 43, 4972.

[9] Osman, MA; Mittal, V; Morbidelli, M; Suter, UW. *Macromolecules*, 2004, 37, 7250.

[10] Osman, MA; Ploetze, M; Suter, UW. J. *Mater. Chem*., 2003, 13, 2359.

[11] Messersmith, PB; Giannelis, EP. *Chem. Mater,* 994, 6, 1719.

[12] Osman, MA; Atallah, A; Suter, UW. *Polymer*, 2004, 45, 1177.

In: Polymer Nanocomposites: Advances in Filler Surface
Editor: Vikas Mittal
ISBN: 978-1-60876-125-8


*Chapter IX*

# Advances in Surface Functionalization of Carbon Nanotubes

***Vikas Mittal****

Institute of Chemical and Bioengineering, Department of Chemistry and Applied Biosciences, ETH Zurich, 8093 Zurich, Switzerland.

## Abstract

Carbon nanotubes, owing to their remarkable mechanical, thermal and electronic properties, have high potential of application in a number of areas as biosensing devices, field effect transistors, chemical sensors, high performance nanocomposites and other nanoscale devices and circuits. These applications require the nanoscale dispersion of nanotubes in the polymer matrices or various solvents. However, nanotubes being inert at surface do not swell in most of the solvents and polymers in their pristine state thus hindering their large scale application. Tremendous research efforts have been undertaken to achieve compatibility of the nanotubes with these media so as to achieve the nanoscale dispersion of the nanotubes which would translate into significantly better and novel properties than any of the existing systems. Two different approaches of covalent and noncovalent modification of nanotubes have been reported to achieve such compatibility. Noncovalent functionalization is based on a number of phenomena e.g. wrapping of polymer chains around the nanotubes, physical adsorption of polymer chains or surfactants on the nanotube surface and deposition of functionalizing molecules by pi-pi stacking etc. Apart from that, charged inorganic particles have been used to achieve the swelling of nanotubes. Generation of thin films of silica or various polymers from the non-covalently bound initiators or monomers have also been reported. The chemical groups generated during the functionalization process were also reacted with proteins to immobilize them on the nanotube surfaces. Noncovalent functionalization of nanotubes leads to nanotube modifications without any loss in the physical properties of nanotubes, however the strength of the modification attached to the nanotubes for load transfer may not be enough. Covalent functionalization has been reported to generate more stronger

* Corresponding author: BASF SE, Polymer Research, Ludwigshafen, Germany. vikas.mittal@basf.com

attachment of the modification on the nanotube surface and thus can be more efficient in load transfer applications. However, the covalent treatment of nanotubes may generate defects in their walls leading to reduction in the mechanical and conducting properties of nanotubes. A large number of methods have been reported for covalent modification of nanotubes e.g. attachment of polyethyleneimine, polyethylene glycol or proteins with amino end groups to the nanotubes, grafting of polymers to the nanotube surface, grafting of polymers from the nanotube surface and plasma treatment for polymer grafting etc. Covalent binding of inorganic nanoparticles to the nanotube surface has also been reported. Use of carbanions to open the pi bonds on nanotube surfaces and subsequent initiation of polymerization from the surface has also been successful. Moreover, various covalent functionalizing techniques avoiding the use of harsh acid treatment of nanotubes have been developed.

**Keywords**: nanotubes, covalent functionalization, noncovalent functionalization, swelling and solubility, grafting to surface, grafting from surface, nanohybrid structures.

## 9.1. Introduction

Carbon nanotubes are allotropes of carbon. Owing to their extremely high physical properties and application potential, nanotubes have been equated to a revolution and have been foreseen as the major components of next generation composite materials [1]. The properties of carbon nanotubes include high modulus and stiffness, electrical conductivity, thermal stability and high aspect ratio and these characteristics make these materials a very attractive choice for high end polymer composites which can achieve the required enhancements in composite properties at very low volume fractions of inorganic materials if nanotubes can be dispersed at nanoscale. Apart from polymer composites with high mechanical and thermal performance, these materials also represent promising class of nanostructured materials for other diverse applications like molecular electronics, chemical sensing, gas storage, field-emission, scanning microscopy and catalysis etc. [2-4]. The credit for realizing the nanotubes in an arc discharge apparatus is given generally to Iijima who could successfully prove the existence of multi walled carbon nanotubes (MWCNT) mixed with other forms of carbon [5]. Though the existence of these materials was realized earlier also e.g. by Endo in 1976, however, no attention was paid to them [6]. MWCNT were classified as either graphite sheets arranged in concentric seamless cylinders or a single sheet of graphite rolled in around itself. Outer diameters from 4-50 nm were reported and length of these tubes was also up to few microns, whereas inner tube diameters as small as 2.2 nm were reported. A tremendous activity in the file followed the initial discoveries and nanotubes with many different methods were generated and characterized [7,8]. Single wall carbon nanocomposites (SWCNT) were later discovered in which one-atom thick sheet of graphite is rolled up into a seamless cylinder. Diameters of the order of a nanometer and length of the order of 1000 nm were achieved depending on the preparation and purification methods. The graphite sheet may be rolled in different orientations along any two-dimensional lattice vector (m,n) defined as helicity or chirality of nanotubes [9].

A number of different methods have been reported in the literature to generate both single walled and multiwalled nanotubes. The prominent among these include arc discharge method [10-12], laser ablation [13], chemical vapor deposition [14-16] etc. In the arc discharge method, carbon contained in the negative electrode sublimates because of the high temperatures caused by the discharge. The carbon nanotubes need to be separated from the soot and the catalytic metals present in the crude product. Depending on the variation of the parameters (e.g. temperature, pressure, different gases and catalytic metals) employed in this technique; it is possible to selectively grow SWCNTs or MWCNTs. In the laser ablation method, pulsed laser vaporizes a graphite target in a high temperature reactor while an inert gas is bled into the chamber. The nanotubes develop on the cooler surfaces of the reactor, as the vaporized carbon condenses. This method provides higher yields up to 70%. In chemical vapor deposition method, a substrate is prepared with a layer of metal catalyst particles, most commonly nickel, cobalt, iron, or a combination. The diameters of the nanotubes that are to be grown are related to the size of the metal particles. The substrate is heated to approximately 700°C. To initiate the growth of nanotubes, two gases are bled into the reactor: a process gas (such as ammonia, nitrogen, hydrogen, etc.) and a carbon-containing gas (such as acetylene, ethylene, ethanol, methane, etc.). The carbon vapor deposition method generates entangled nanotubes in general; however, more aligned tubes can be generated by using the synthetic condition which leads to rapid and dense nucleation on the substrates [17-18]. The major differences between the synthesis techniques are considerations like yield, quality, scale up and commercialization as the nanotubes still are an expensive filler.

The potential of nanotubes' application stems from their remarkable properties which single them out from other fillers or inorganic materials. They have very high aspect ratios and chemical bonding in these tubes is composed entirely of sp2 bonds, similar to those of graphite. Nanotubes naturally align themselves into so called ropes held together by van der waals Waals forces. They have remarkable mechanical properties with tensile strength estimated to be around 63 GPa and elastic moduli in the range of 1 TPa [19-24]. Another important property is related to their relatively high strain bearing capacity as plastic deformation is observed only after 5% strain. CNTs can either be conducting or semiconducting, depending on their chirality, thus covering the whole spectrum of electrical conductivities. Nanotubes are extremely stable even at very high temperatures (>2500°C) in vacuum [25]. Multi walled nanotubes have special wider application potential because they improve significantly the resistance to chemicals especially when chemical functionalization of the nanotube surface is required.

A number of potential applications of nanotubes require their dispersion in polymers and solvents. In order to generate high property nanocomposites, the nanoscale dispersion of nanotubes in the polymer matrices is of utmost importance to generate organic inorganic interface which has altogether different morphology than the individual components. Also, the nanotubes should be dispersible in the polymer matrices when processed either in melt, solvent solution or during in situ polymerization. The nanotubes, however, are chemically inert and form bundles because of Van van der Waals forces between them. This hinders their nanoscale dispersion in polymer matrices when added in the pristine state indicting that the improvements in the interfacial interactions are required for the optimum enhancement in the composite properties. The functionalization on the surface makes the nanotubes more organophilic and hence helps to delaminate inside the polymer matrices by the application of mechanical shear. To circumvent the limitation of incompatibility of the polymer and nanotubes, a number of methods have been reported to functionalize the nanotube surface to achieve compatibility. Prominent among these are non-covalent functionalization and covalent functionalization methods. Non-covalent functionalization methods include polymer wrapping, adsorption of surfactants on the nanotubes surface and pi-pi stacking etc. The covalent functionalization methods include grafting of polymer chains to or from the surface of nanotubes. Both methods have their own advantages and limitations, however, both suggest very attractive routes to ease the use of nanotubes for commercial applications. Some other methods have also been described in literature e. g. dispersion of nanotubes in polymer solutions by using high energy sonication.

## 9.2. Non-Covalent Functionalization of Nanotubes

This method is based on physical adsorption of polymer chains on or wrapping of polymer chains around the nanotubes surface. Advantages of these methods are that the structure of nanotubes remains unaltered and the mechanical properties are not reduced. However, it suffers from the limitation that the forces between the wrapping molecule and nanotube may be weak which leads to low load transfer efficiency in the system. Many studies on this mode of functionalization have been studied and have been discussed in the following paragraphs [25-59]. Figure 1 also explains these approaches pictorially.

O´Connell et al. reported polymer wrapping method for dispersion of nanotubes in water [26]. SWCNTs were dispersed in 1% sodium dodecyl sulfate (SDS) in water by using shear mixer and ultrasonication. Poly(vinyl pyrrolidone) (PVP) dissolved in water was then added to this suspension of nanotubes in excess. The excess PVP was subsequently removed by ultracentrifugation, decanting the serum and rewashing with water steps. The suspensions were observed to be stable for months and once dried, the nanotubes were reported to disperse in water with minimal ultrasonication. Non-wrapped nanotubes were observed to assemble into mats of tangled and seemingly endless ropes, whereas it was not the case in the polymer wrapped nanotubes. It was hypothesized by the authors that the nanotube wrapping is driven largely by the thermodynamic drive to eliminate the hydrophobic interface between the tubes and the dispersion medium. It was also hypothesized that the polymer wrapping around the nanotubes took place by following multi-helical wrapping of the polymer chains because this way allows high surface area coverage with low back bone strain.

Polyvinyl alcohol (PVOH) was coated around the nanotubes owing to its hydrophilic nature thus allowing the dispersion of nanotubes in water [27]. The nanotubes were submerged in the aqueous solution of PVOH which was hydrolyzed to 99%. It was clearly observed that the deposition of even a small amount of polymer on the nanotube surface changed the morphology of the nanotubes significantly. Dynamic contact angles measured on the surface of the nanotubes indicated a marked decrease from 148° to 24° when the nanotubes were functionalized on the surface with PVOH confirming that the polymer modification led to significant reduction in the surface energy of the nanotubes. It was also hypothesized that PVOH being semi crystalline in nature, crystallizes and immobilizes alcohol functional group on the surface of the tubes, which can be used for further processing with other molecules.

Similarly, conjugated luminescent polymer poly(metaphenylenevinylene) (PmPv) was coated on to the nanotubes surface [28]. It was suggested that high molecular weight of the polymer is more efficient in disrupting the van der Waals forces between the nanotubes which otherwise cause them to arrange in bundles. Spectroscopically, it was confirmed that PmPV polymer wrapped itself around the SWCNTs, in a way that the phenylene rings and vinyl units from the polymer backbone came close to the surface of nanotubes leaving the octyloxy chains attached to the polymer backbone to wiggle around in solution, It was further confirmed by AFM that polymer entered into significant noncovalent bonding with the nanotubes. However, it was also observed that the polymer tended to wrap around the nanotube bundles than the single nanotube itself indicating processing improvements are required along with the suitable characterization methods to monitor the functionalization process. In an another study, synthesis and characterization of polymers functionalized in C5 position on the meta-disubstituted phenylene ring (poly{(5-alkoxy-m-phenylenevinylene)-co-[(2,5-dioctyloxy-p-phenylene)-vinylene]}) was reported [29]. These polymers had tethers or rings that form pseudorotaxanes with matching rings or threads. These polymer chains were successfully wrapped around the carbon nanotubes which resulted in the grafting of pseudorotaxanes along the walls of the nanotubes.

An interesting approach of noncovalent functionalization of the nanotubes was suggested by Zhu et al. in which charged inorganic nanoparticles were used as stabilizing media [30]. Single wall carbon nanotubes were mixed with $ZrO_2$ nanoparticle solutions and were sonicated. After initial sedimentation of the uncoated nanotubes, the suspensions of nanotubes with nanoparticles were observed to be very stable for long periods of time. AFM studies on the samples also showed the existence of single nanotubes in the dispersed mode indicating that the nanotubes were protected from aggregating with each other by the use of Zirconium zirconium dioxide nanoparticles. The authors suggested that the most plausible stabilization mechanism was based on charge repulsion originating from Zirconium zirconium dioxide particles. The external stimuli were observed to self assemble the dispersed nanotubes into macroscopic materials in solution indicating that the functionalization of the nanotubes with the nanoparticles was very efficient.

Stoddart et. al. reported an interesting study on the starched carbon nanotubes in which the nanotubes were wrapped with starch [31]. Starch was first activated towards complexation by wrapping itself around small molecules. It was observed that the SWCNTs were soluble in this starch iodine complex, whereas they were insoluble in aqueous solution of starch itself. It was suggested that amylase in starch was organized into a helical conformation by complexation with iodine and it eases the way for a single carbon nanotube or bundles by replacing the small iodine molecules inside the helix, whereby enthalpic gain from favorable van der Waals interactions is aided and abetted by entropy driven process. Subsequently, addition of glycosidase to the starched carbon nanotubes resulted in the precipitation of the tubes from aqueous solution.

Physical adsorption of long chain polymers was reported by Regev et al. to disperse the nanotubes individually in solvents [32]. Non-specific adsorption of natural polymer, Gum Arabic (a polysaccharide), was achieved by dispersion of SWCNTs in the aqueous solution of Gum Arabic. The dispersion was then dried and was observed to disperse subsequently well in high concentrations in water. By microscopic investigations, it was confirmed that the adsorption of Gum Arabic led to the disruption of the internal structure of the crystalline ropes. Entropic repulsion among polymer chains was thus successfully utilized for stabilization of nanotubes, the phenomena also called steric stabilization.

In an interesting study on the interactions of proteins with nanotubes, it was demonstrated that by noncovalent functionalization of nanotubes, non- specific binding of the proteins on the nanotube surface could be eliminated and specific binding of proteins after surface functionalization could be achieved [33]. It was observed in the study the poly(ethylenegylcol) (PEG) which is commonly used fror polymer wrapping [34-37], though irreversibly adsorbed on the nanotube surface, but appreciable adsorption of streptavidin was still observed indicating that the coverage of nanotube with PEG was not optimum. To circumvent the problem, a special surfactant was initially used to coat the nanotubes which significantly enhanced the wrapping of nanotubes with PEG confirmed by the observation that no streptavidin adsorption was observed after surface functionalization. For specific binding of proteins on the functionalized nanotubes, a biotin moiety was added on the PEG chains by using amine functionalized PEG and covalently linking it with an amine reactive biotin reagent. Very specific adsorption of streptavidin was observed on these triton-PEG-biotin functionalized nanotubes.

An inorganic silica layer was generated around the carbon nanotubes by following the noncovalent functionalization methods [38,39]. A promoter, 3-aminopropyltriethanoxysilane (APTES) was absorbed on the sidewalls of the nanotubes by the interaction of side walls of nanotubes with the amine groups. The molecule was subsequently polymerized by heating and thus irreversible coating on the nanotube surfaces was generated. As a next step, thin shell of $SiO_2$ was generated around the nanotubes, the thickness of which could be controlled by adjusting the growth conditions. This way, an inorganic $SiO_2$ layer could be generated which was non-covalently bound to the nanotubes' surfaces. This can then further allow using the $SiO_2$ functionalization chemistries to generate different functionalities on the $SiO_2$ surface, still keeping the nanotube core mechanically intact. It thus leads to synergistic equilibrium between the nanotube core and silica shell in which properties of both the components can be positively utilized. Similarly, norbornene polymerization was achieved on the nanotube surface via non-covalently bound ring opening metathesis initiator [40]. Single wall carbon nanotubes were non-covalently functionalized with ruthenium alkylidenes by using various synthetic routes. These alkylidenes were used for ring opening polymerization process to generate a thin coating of polynorbornene on the nanotube surface. Varying thicknesses of the polynorbornene shell were formed by using different synthetic methods and reaction conditions. As earlier reported technique, this technique also serves to generate polymer grafts from the nanotube surface without chemical functionalization of the nanotubes, thus, retaining the mechanical performance of the tubes.

Polymer multilayers were formed on the non-covalently modified nanotubes in an interesting study by Kane and Ajayan et al. [41]. Formation of multilayers have been has been reported in a number of earlier studies also to generate functionalized surfaces [42-51]. In this study, hydrolyzed-poly(styrene-alt-maleic anhydride) (h-PSMA) was adsorbed non-covalently onto nanotubes surface via hydrophobic interactions. As the copolymer contained carboxylic acid groups, therefore, these groups could be used to further attach a second layer of polyethyleneimine covalently. The same process was followed to attach second layer of h-PSMA and subsequently a second layer of polyethyleneimine. In the end, a layer of polyacrylic acid was generated on the surface of the nanotubes. The authors this way could confirm the versatility of the process to generate multi layers on the nanotube surface. Gold nanoparticles could also be immobilized on the surface by generating a third layer of polyethyleneimine. As gold particles were negatively charged, they could bind to the positive charges in the polyethyleneimine layer by electrostatic interactions. The microscopic as well as spectroscopic studies could be successfully employed to monitor the various processes involved in the functionalization cycle.

Encapsulation of polymer nanotubes within cross-linked, amphiphilic copolymer micelles was reported by Taton et al. and other researchers [52-54]. This novel process of nanotube encapsulation was observed to be significantly beneficial in enhancing the dispersion of nanotubes in solvents and polymer matrices. Amphiphilic poly(styrene)-block-poly(acrylic acid) copolymer was dissolved in dimethylformamide (DMF) followed by the suspension of nanotubes by ultrasonication. Gradual addition of water to the suspension led to the micellization of the copolymer with encapsulated nanotubes inside these micelles. The poly(acrylic acid) blocks were subsequently crosslinked by the addition of diamine linker and carbodiimide activator. It was also observed that full crosslinking of poly(acrylic acid) blocks was not necessary to achieve stabilization, conversion of 25% acid groups in amides was enough to bring about the stability of nanotube suspension. The encapsulated nanotubes could be washed and dried followed by successful redispersion in a variety of hydrophobic and hydrophilic solvents and polymer solutions. Apart from that, these suspensions were stable over long periods of time. It was hypothesized that high solubility of these encapsulated nanotubes resulted because of solvation of at least one block of the encapsulating copolymer in the solvents used.

Non-covalent functionalization of carbon nanotubes' was reported by using aromatic polyimides as functionalizing polymer [55]. Along with the high end properties of polyimides, it was also expected that an effective pi-pi stacking can form between the polyimide and the graphitic side walls of the CNT owing to the pi electron system in imide and benzene ring of polyimide. In this way, the precursor of polyimide could be non-covalently attached to the nanotube surface which was then treated to form chemically bound polyimide. Infra Rred, Raman and UV-visible spectroscopic methods confirm the presence of pi-pi stacking interaction and the thermal stability of the whole system after functionalization was improved. It was also confirmed through transmission electron microscopy that an amorphous layer of polymer was present on the surface of nanotubes. The diameters of the nanotubes were also confirmed to increase after surface functionalization with polyimide.

Further trials for the immobilization of biomeolecules on carbon nanotubes were reported by Chen et al [56]. Non-covalent functionalization of the sidewalls of single walled carbon nanotubes was achieved and subsequent immobilization of various biological molecules was observed and the process was reported to be controllable and site specific. A bifunctional molecule, 1-pyrenebutanoic acid, succinimidyl ester was adsorbed on the basal plane of nanotubes via pi stacking [56-58]. It was also observed that the ester molecule was irreversibly adsorbed on the surface and resisted desorption in aqueous solutions. The protein immobilization on the surface could then be achieved by nucleophilic substitution on N-hydroxy succinimide by amine groups on the protein. Successful immobilization of both ferritin and streptavidin proteins could be achieved by this method thus indicating the versatility of this functionalization approach to make the surfaces of nanotubes more usable. Apart from proteins, any other chemical molecules can also be reacted on the surfaces treated with ester and thus the surface properties can be tuned or controlled according to the requirement.

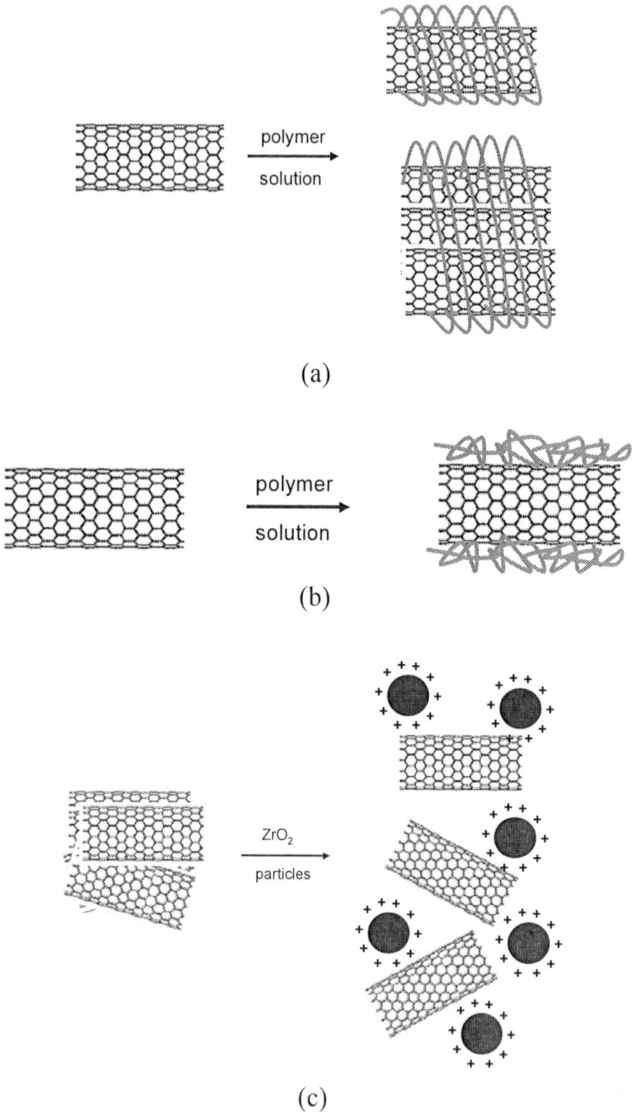


Figure 1. Various methods of noncovalent functionalization of nanotubes; (a) polymer wrapping approach [26-29], (b) physical adsorption of polymers [32], (c) use of charged inorganic nanoparticles [30].

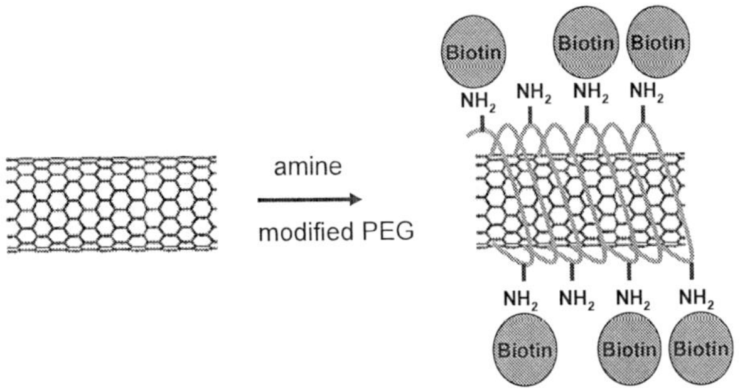


(d)

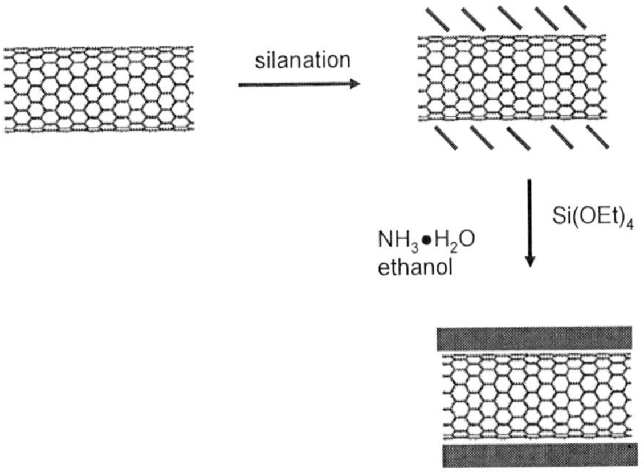


(e)

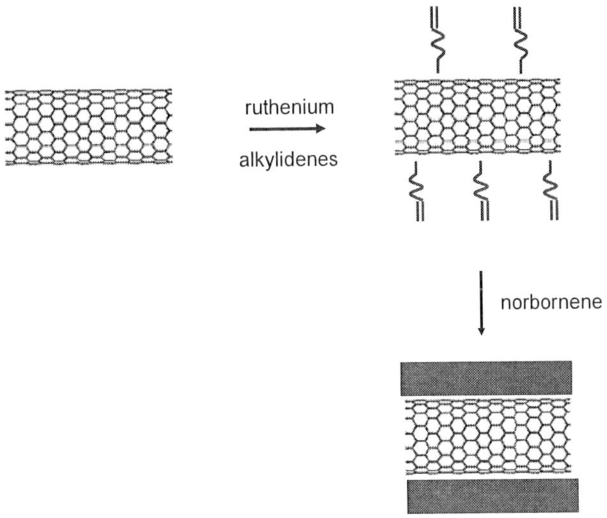


(f)

Figure 1 (contd.). Various methods of noncovalent functionalization of nanotubes; (d) use of $NH_2$ containing polymer chains to bound proteins [33], (e) generation of thin silica layer around the nanotubes [38,39], (f) polymerization of norbornene around nanotubes [40].

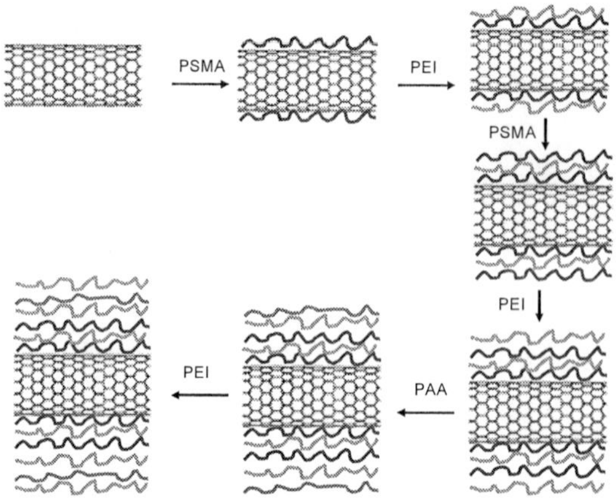


(g)

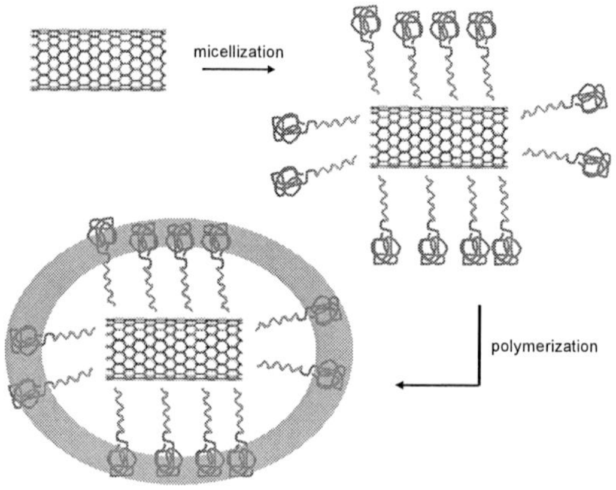


(h)

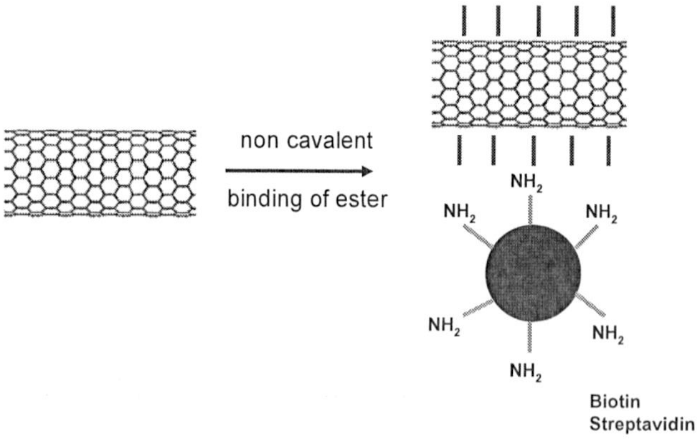


(i)

Figure 1 (contd.). Various methods of noncovalent functionalization of nanotubes; (g) use of polymer multilayers [41], (h) encapsulation of nanotubes [52], and (i) pi-pi stacking of reactive ester molecule to chemically bind proteins [56].

In another interesting study, a graft copolymer polystyrene-g-(glycidyl methacrylate-co-styrene) (PS-g-(GMA-co-St)) was used to non-covalently modify the surface of nanotubes [59]. It was proposed by the authors that the adsorption of polymer chains is based on the affinity of polystyrene chains to the surface pf of pristine nanotubes. Interestingly, the same functionalizing copolymer was observed to behave differently when nanotubes were acid treated. In this case, the acid groups present on the nanotube surfaces reacted with the gylcidyl methacrylate component of the copolymer and a covalent binding of the copolymer to nanotube surface took place. It was observed that different modification mechanisms significantly affected the solubility characteristics of the nanotubes. The nanotubes modified non-covalently were observed to swell well in different polar and non-polar solvents, whereas the covalently functionalized nanotubes were swollen in solvents in which polystyrene is soluble. The covalent functionalized nanotubes had polystyrene chains as external 'capsules' owing to the chemical reaction of the other block of the copolymer with the surface of nanotubes. Thus, the swellability of nanotubes was dictated by the swelling or solubility characteristics of polystyrene chains. However, in the case of non-covalently functionalized nanotubes, the polystyrene component of the copolymer was attached to the nanotubes surface this thus leaving the chains containing gylcidyl methacrylate dangling away form the nanotube surface and forming the outside 'capsule'. As glycidyl methacrylate is readily soluble in many solvents, thus the swelling of non-covalently functionalized nanotubes was improved significantly. ThusTherefore, it is clear from this study that the nature of same surface functionalization can also be changed by changing the nature of substrate surface and these modifications thus can be finely controlled according to requirement.

## 9.3. Covalent Functionalization of Nanotubes

This method is focused towards grafting of polymer chains bound chemically to the nanotube surface. It thus eliminates the limitation of the non-covalent functionalization system of poor load transfer between the polymer and nanotube. However, the chemical functionalization may introduce defects on the walls of nanotubes by disrupting the extended pi-networks on their surfaces, which may lower both the electronic and mechanical properties of the reinforcing component. The choice of the functionalization method is best governed by the requirements of the system and intended applications of functionalized nanotubes. Numerous studies focused on the subject of covalent functionalization of nanotubes which are mentioned as follows [60-127]. Figure 2 shows these approaches pictorially.

A covalent functionalization of nanotubes with polyethylenimine (PEI) was reported in order to achieve immobilization of DNA on the functionalized nanotubes [60]. Multi walled carbon nanotubes were heated at reflux with nitric acid which generated carboxylic acid groups on the surface of nanotubes. These acid groups were transformed into acyl chloride groups by treatment with thionlyl chloride followed by treatment with ethylene diamine to result nanotubes with $NH_2$ groups on the surface. Polyethylenimine was grafted from the surface either by attachment of protonated aziridine monomers to amine groups on nanotubes' surface or terminal iminium ion groups of propagating chains were transferred to these amine groups. Amount of grafted PEI could be thermo-gravimetrically quantified and the modified nanotubes were observed to swell readily in the solvents. It was observed spectroscopically that a small percentage of amine groups in PEI chains were protonated and this protonation was necessary to achieve stable aqueous suspension of PEI grafted nanotubes and the stability of the suspension was totally disturbed when PEI chains were neutralized by adjusting the pH to 9 or higher.

Functionalization of single walled carbon nanotubes with enzymes was reported following the similar covalent modification route as above [61] as compared to other non-covalent modification approaches [62-65]. Acylation of SWCNT was followed by amidation with either enzyme or any amine indicating the versatility of the modification method. Successful reactions of the following amine containing molecules with the surface groups of nanotubes could be achieved: porcine pancrease lipase (PPL), amino lipase (AK), cis-myrtanylamine, 2,4-dinitroaniline, 2,6-dinitroaniline, N-decyl-2,4 6-trinitroaniline and N-(3-morpholinopropyl)-2,4,6-trinitroaniline). It was observed that the generated defects and covalent binding of amines on the surface after functionalization caused the reduction of electrical conductivity in the functionalized nanotubes. It was also observed that both the single nanotubes as well as bundles could be grafted indicating that initial homogenous dispersion of nanotubes during functionalization is immensely important. Successful immobilization of a variety of amine as well as biological entities on the surface confirmed the potential as well as robustness of the functionalization method. In a similar study, diamine terminated oligomeric poly(ethylene glycol), O,O'-bis(3-aminopropyl)poly(ethylene glycol) was reacted with acid treated SWCNTs at 100°C [66]. In another approach, carbodiimide was employed as activator to achieve covalent functionalization of the nanotubes. For the carbodiimide activated amidation, acid treated SWCNTs were suspended in $KH_2PO_4$ buffer followed by the addition of polyethylene glycol and 1-ethyl-3-(3-dimethylaminopropyl) carbodiimide. Thus, a number of different permutations with the synthetic methods as well as reacting amines can be used.

Poly(methyl methacrylate) (PMMA) was polymerized in- situ in the presence of un-functionalized and acid functionalized nanotubes and interesting phenomena regarding the mode of polymerization and comparison of the resulting composite properties were reported [67]. It has been suggested in earlier studies too that the initiator can open the pi bonds found in pristine nanotubes and subsequently nanotubes can take part in polymerization reaction, thus grafting the polymer chains covalently bound to the clay surface [68,69]. However, this pathway was reported to be less attractive than the covalent functionalization of nanotube which can then be used to graft a variety of molecules on the nanotube surface. Owing to PMMA polymerized in the presence of unmodified nanotube, the position of vibration bands in the IR spectra was observed to increase indicating pi bonds in the nanotubes could have been opened. Composites generated using functionalized nanotubes were confirmed through IR spectroscopic characterization to have a chemical connection between the formed polymer and nanotubes. Composites generated by both unmodified and unmodified nanotubes though had higher mechanical properties than the pure polymer and the properties were more significantly enhanced when the nanotubes were modified first at the surface. Thus, it was suggested that covalent surface functionalization in presence of in -situ polymerization to attach the polymer chains covalently to the surface of nanotubes is the best method for generating nanocomposites which represent synergistic performance of the components.

Generation of polymer brushes of poly(n-butyl methacrylate) with defined properties was achieved from the surface of nanotubes by using controlled living polymerization method [70]. In a polymer brush the chains are immobilized at one end on the substrate, whereas the other end is free for achieving chemical reactions [71-76]. If these brushes are generated by using living polymerization techniques, the living agent can also allow the subsequent formation of block copolymers grafted from the surface of nanotubes, thus controlling the surface characteristics of the nanotubes. Grafting from the surface methods have been studied widely in detail and can be directly applied on the nanotubes [77-82]. In this study, nanotubes were refluxed in nitric acid to generate acid functional groups on the side walls of the tubes which were then reacted with thionyl chloride to convert them into acyl chloride groups. These acyl chloride end groups were reacted with and atom transfer radical polymerization agent 2-hydroxyethyl-2-bromopropionate to generate SWCNT modified with ATRP initiator. These nanotubes could then be used to graft brushes of poly(n-butyl methacrylate) in the presence of CuCl and BiPy. These brushes were confirmed to have significant effect on the solubilization of nanotubes. The degree of grafting could also be controlled during the polymerization process and by using ATRP initiator bound to the surface, a number of different polymers could be grafted from the surface thus providing flexibility in the functionalization process. Gao et al. also reported the grafting of poly(methyl methacrylate) brushes from the surface of nanotubes by using similar grafting from surface approach [83]. In this study, carbonyl chloride groups on the surface of nanotubes were reacted with glycol to generate MWCNT-OH, which were subsequently reacted with ATRP initiator 2-bromo-2-methylpropionyl bromide. Good swellability of poly(methyl methacrylate) functionalized nanotubes in non- polar or weakly polar solvents was observed. Polymer thicknesses from 3.8 nm to 14 nm were grafted by changing the feed ratios. The brushes were also subjected to further polymerization of hydroxyethyl methacrylate (HEMA) using poly(methyl methacrylate) functionalized nanotubes as macroinitiator indicating copolymers with different vinyl polymers can be achieved to change the surface nature of nanotubes according to the requirement.

Plasma deposition of polymers as thin films on the surface of nanotubes has also been reported [84]. The resulting nanotubes were then more combatable compatible with the solvents and polymer matrices in which they were subsequently compounded. The nanotubes were treated with plasma in the plasma chamber followed by the introduction of monomers at high pressure. The monomers could also be labeled partially with fluorine to characterize the deposited films in time of flight secondary ion mass spectrometry. It was observed that thin films of 2-7 nm thickness of polypyrrole could be deposited on the outer and inner surfaces of nanotubes and the deposited films were observed to be uniform. The thin films contained highly branched and crosslinked polymer chains. It was remarked that by controlling the plasma deposition conditions, the thickness of the inner and outer films can be efficiently controlled.

In another study of covalent functionalization of nanotubes to attach functional molecules on the surface, attachment of diglycidyl ether of bisphenol A on the nanotubes surfaces was studied [85]. MWCNT were first modified to generate surface acidic groups and were subsequently reacted with epoxide terminated molecule. The use of bifunctional molecule for attachment on the nanotube' surface leads to subsequent functionalization reactions from the free end. It was observed from the IR spectrometry that both the epoxide groups in the functionalizing molecule were reacted either by the epoxide-epoxide reactions that follows the reaction of the epoxide ring with the carboxylic acid group on the nanotube, or both the epoxide groups can react with the acid groups present on the nanotube surface. It was suggested that an epoxy polymer with high molecular weight can also be used to form polymer composites in which epoxy matrix would be chemically bound to the nanotube surface. It was also remarked that in the case of thermoplastic bulk polymers, the van der Waals forces between the functional groups and the nanotubes define the load transfer mechanisms. It was suggested to use long enough surface modifications which can then entangles with the polymer chains at interface leading to better load transfer. The authors also reported the generation of polycarbonate nanocomposites in which the polycarbonate chains could be tethered to epoxy functionalized nanotubes by transesterification reactions [86-89]. The large available surface area of the nanotubes results in significant amounts of polymer immobilization on the surface which could also be controlled by changing the surface modification of the nanotubes. The enhanced intermixing caused by the epoxy functionalized nanotubes led to improved mechanical, thermal and load transfer properties as compared to the composites with non-functionalized nanotubes. This clearly indicates that the modification of nanotubes with functional molecules holds the key to successful generation of composites where nanotubes are dispersed on nanoscale and the properties are synergistically enhanced.

Few studies have also focused on the modeling of the nanotube functionalization by polymer grafting [90-98]. In such an outstanding study by using the density functional theory, possible ways of functionalization of carbon nanotubes using methoxy radical and secondary butyl anion as initiators were studied followed by their reactions with ethylene and epoxide, leading to polyethylene- and polyepoxide grafted nanotubes [90]. The study predicted that both free-radical and anionic functionalization methods were energetically favorable and the resulting CNT radical and the CNT anion could react with ethylene and epoxide, respectively. The resulting products were observed to have free electrons and negative charges on the carbon and oxygen atoms at the free ends of ethylene and epoxide, respectively. However, the cationic functionalization of carbon nanotubes using $BF_3$ as an initiator was found to be infeasible.

Grafting to the surface approach has been reported to generate high density grafts on the surface, as compared to grafting to approach where the polymer chains are grafted onto the nanotube surface by direct esterification or amidation reaction. Grafting to the surface is also achieved when by in- situ polymerization of monomers in the presence of nanotubes modified with reactive groups. Owing to the steric repulsion in the case of grafting to surface method, linear polymer chains are easier to graft. In a study to underline these phenomena, acid groups on nanotubes surfaces were reacted with poly(acryloyl chloride) [99]. Subsequently, these acryloyl chloride groups were reacted with linear polyethylene glycol to generate grafts attached to the nanotube surfaces. Significant amounts of polymer could be grafted to the surface indicating concerns regarding steric repulsion could be eliminated in the technique.

To gain application in biotechnology, immunosensing and biomedical applications, attachment of quantum dots on the surface of nanotubes was studied, which can be used to specifically immobilize biomelecules biomolecules on the surface [100-104]. CdSe nanoparticles were attached with mercaptoacetic acid ligands which were reacted with amine groups present on the nanotube surface. These groups could be generated by reacting COCl groups on the nanotube surface (achieved by acid and thionyl chloride treatments) with ethylene diamine. By forming the similar amide bonds, high performance diamine functionalized polyetheri-ides could be covalently immobilized by reaction of the amine end groups with the acid groups present on the nanotube surfaces [105].

In yet another interesting study, a novel approach to in- situ composite synthesis by attachment of polystyrene chains to SWCNTs without disrupting the original structure, based on an established anionic polymerization scheme was reported [106]. The reported process required no nanotube pretreatment and both suspension of SWCNT and polymer grafting were achieved in a single step. The process resulted into well-defined composites with a homogeneous dispersion of nanotubes. To achieve the grafting, SWCNTs were dispersed by sonication in purified cyclohexane to which a slight excess of butyl lithium was added. This introduced carbanions on the SWCNT surface which help to exfoliate the nanotube bundles because of the repulsion between the negatively charged nanotubes and also provided initiating sites for the polymerization of styrene. On addition of styrene, both free butyl lithium and the nanotube carbanions were observed to initiate polymerization, resulting in an intimately mixed composite system. The polymerization could efficiently be terminated by the addition of degassed *n*-butanol.

Fluorination of SWCNTs has been one of the first covalent functionalization approaches to generate single nanotubes compatible with solvents and polymers [107-109]. It was also reported in a number of studies that fluorinated SWCNTs could be further derivatized due to higher reactivity than the un-functionalized nanotubes [110-112]. In one of such studies, a one-step functionalization method using fluorinated single wall carbon nanotubes (F-SWCNT) to react with urea, thiourea, or guanidine was reported [112]. As a result, the derivatives with terminal amide and heteroamide groups on the nanotube sidewalls could be prepared. Urea-functionalized SWCNTs (U-F-SWCNT) were observed to show the highest stability of their dispersions in DMF, water, and aqueous urea solutions, thus confirming their potential use for biomedical applications. Bifunctional derivatives could also be generated during these reactions owing to the presence of residual fluorine in the nanotubes and these derivatives were observed to have improved dispersion properties in the epoxy system indicating the possible use of such functionalized nanotubes to generate polymer composites in which the nanotubes are dispersed at nanoscale.

Kakade et al. reported the use of microwave irradiation for the acid functionalization of both single and multi walled carbon nanotubes which resulted in the generation of significant amount of hydrophilic functional groups on the surface [113]. The acid groups could be converted to COCl groups by using oxalyl chloride and subsequent reaction of these reactive groups with aniline provided a remarkable degree of aniline functionalization through amidation, and the formation of polyaniline could be avoided. Electrochemical analysis confirmed that the variations on the nanotube topography were consistent with the concomitant formation of a dynamic polymer layer. By changing the processing parameters like duration of microwave irradiation, nature of the medium and the input power etc., the extent of functionalization could be varied.

Nanohybrid materials from CNTs and nanoparticles were reported in which the nanoparticles were covalently immobilized on the nanotubes surface [114-117]. Immobilization of nanoparticles on the nanotubes surface has earlier also been reported non-covalently, but the covalent functionalization was reported to be more efficient and stronger in interactions between the components. Magnetite colloid solution was prepared in ethanol solution, to which 3-aminopropyltrimethoxysilane (APTS) was added with rapid stirring. APTS-coated magnetite nanoparticles were dried under vacuum at room temperature which resulted in amine-terminated magnetite nanoparticles. These particles were reacted with COCl treated nanotubes at room temperature using ultrasonification. The successful immobilization of inorganic magnetite nanoparticles indicates that the covalent binding process can be used as a general approach to effective functionalization of carbon nanotubes with various nanoparticles for a wide range of potential applications.

Covalent functionalization of multi walled carbon nanotubes (MWCNTs) was also reported with a low molecular weight chitosan by a nucleophilic substitution reaction on the surface of nanotubes [118]. Amino and primary hydroxyl groups of chitosan were reported to contribute mainly to the formation of MWNT-chitosan structures by reacting with COCl groups generated on the nanotube surfaces by acid treatment. A significant amount of chitosan could be grafted to the surface as its content in the functionalized chitosan-nanotube structures was approximately 58 wt %. It was reported that approximately four molecular chains of chitosan were attached to 1000 carbon atoms of the nanotube sidewalls. Additionally, the nanotubes were observed to induce crystalline character in chitosan. The functionalized nanotubes were soluble in a variety of solvents like dimethylformamide, dimethyl acetamide, dimethylsulfoxide and acetic acid aqueous solution.

Plasma treatment of nanotubes was earlier shown to graft polystyrene on the nanotube surfaces [84]. In yet another study, plasma treatment was used to generate polymer nanotube nanocomposites in which the polymer was chemically bound to the surface of nanotubes [119]. The functionalization of MWCNTs was first achieved via plasma treatment, which led to the generation of a large number of radicals on the surface of the MWCNTs. To these nanotubes, maleic anhydride (MA) dissolved in toluene was added inside the reactor to graft MA onto the MWCNTs. These functionalized nanotubes were then mixed with diamine curing agent followed by a further reaction with the epoxy matrix. It could be observed that the modified CNTs were covalently integrated into the epoxy matrix and became part of the cross-linked structure rather than just a separate component. The modified nanotubes had better dispersion inside the polymer matrix as well as on the fracture surfaces of the composites as compare to unmodified nanotubes.

As acid treatment is a harsh process of functionalization of nanotubes, an alternative method was suggested to covalently modify the nanotubes without using the acid treatment [120-124]. The purified MWCNTs were functionalized with formaldehyde via an electrophilic substitution reaction by dispersing the nanotubes and NaOH in formaldehyde solution. These nanotubes were added with acrylic acid and perchloric acid and the reaction was performed under reflux. Polyacrylic acid (PAA) grafting on the surface of nanotubes could be achieved by reacting acrylic acid treated carbon nanotubes with externally added acrylic acid and free radical initiator. The PAA wrapped around the MWCNTs was observed to improve greatly the hydrophilicity of carbon nanotubes.

The IR and TEM results showed presence of covalent bond and so-called ‘‘core-shell’’ structures for poly(acrylic acid) grafted nanotubes.

Covalent sidewall functionalization of single-walled carbon nanotubes was also reported via one-electron reduction of benzophenone by potassium [125]. The reaction of a potassium atom with a benzophenone molecule resulted in transferring one electron from potassium to benzophenone, forming a radical anion, which could be readily added to nanotubes. It was confirmed spectroscopically that a covalent modification of the nanotubes could take place. In another approach, addition of an acyl peroxide derivative could be readily achieved by directly adding it to a suspension of reduced SWCNT [126].

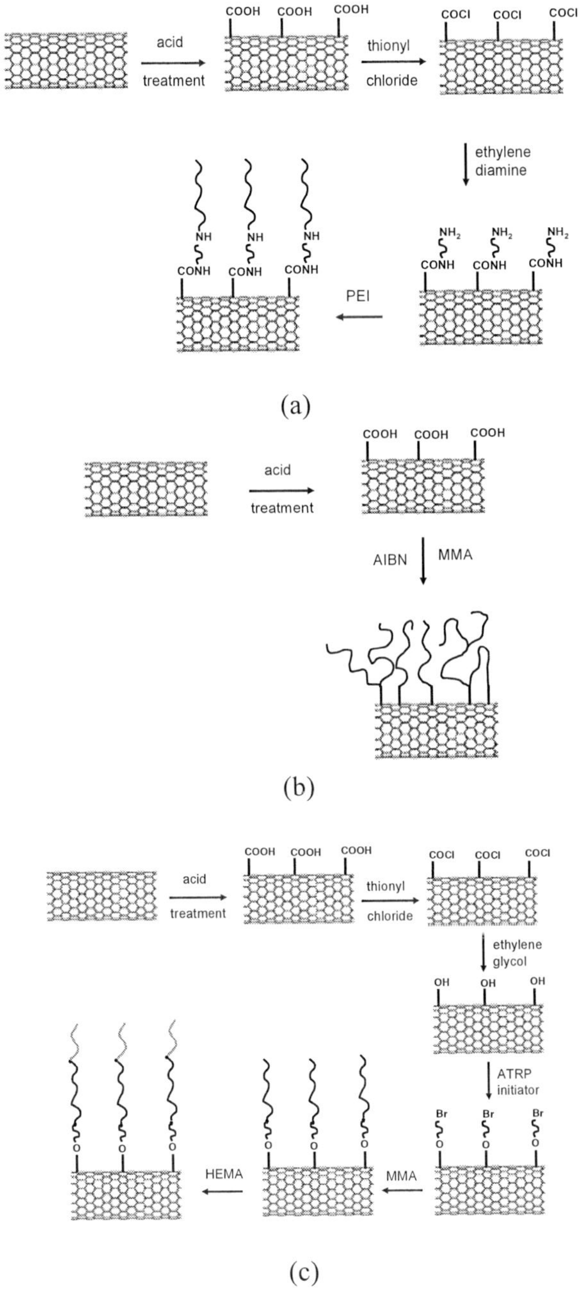


Figure 2. Various methods of covalent functionalization of nanotubes; (a) attachment of polyethyleneimine to the surface [60], (b) grafting to surface approach to graft PMMA [67], (c) grafting from surface approach to graft PMMA [83].

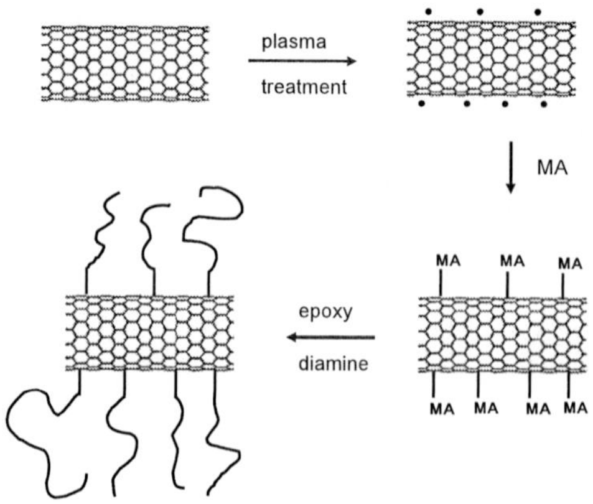


(d)

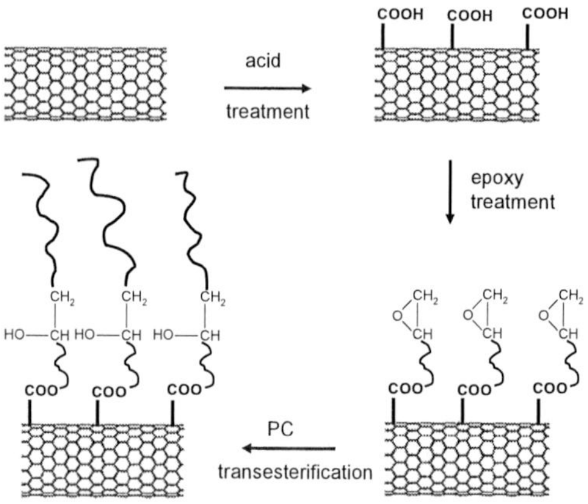


(e)

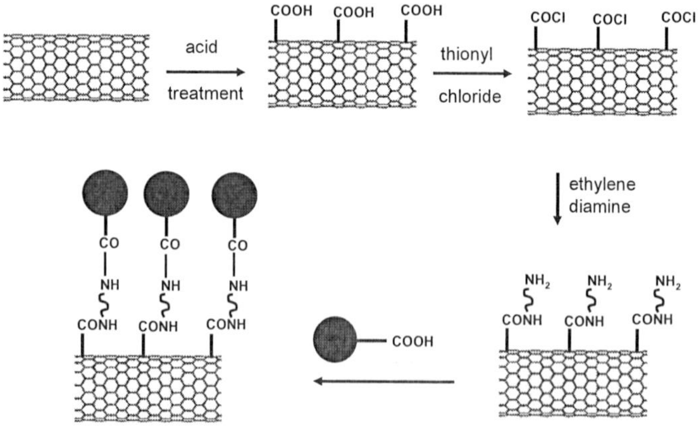


(f)

Figure 2 (contd.). Various methods of covalent functionalization of nanotubes; (d) plasma treatment to graft polymers [84,119], (e) attachment of polymers through epoxide functionalization [86], (f) covalent attachment of quantum dots [105].

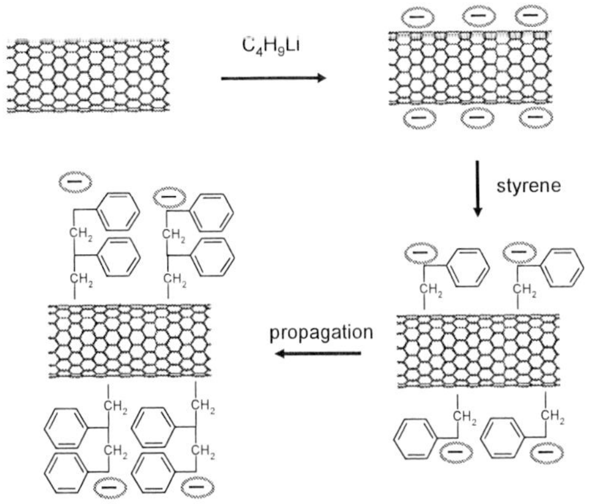


(g)

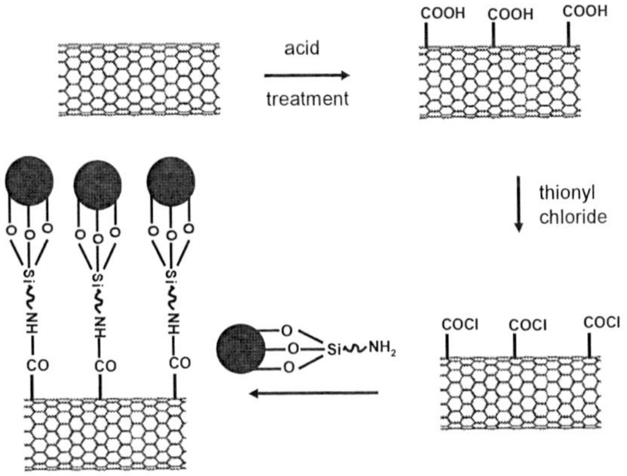


(h)

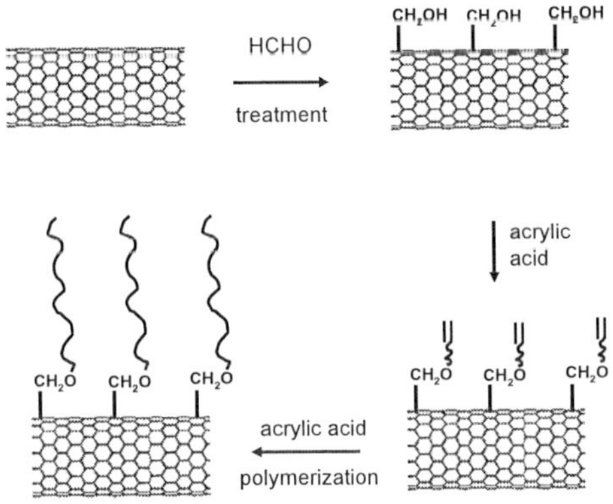


(i)

Figure 2 (contd.). Various methods of covalent functionalization of nanotubes; (g) generation of carbanions and polymer grafting [106], (h) covalent binding of inorganic nanoparticles [114] and (i) modifications avoiding the use of harsh acid treatment [120].

Reduced SWCNT were prepared by electron transfer from alkali-naphthalene salts or alkali-metal benzophenone salts. The acyl peroxide derivatives reported were terminated with alkyl, phenyl and carboxylic acid functionalities, however, it was reported that the method could be compatible with any acyl peroxide derivatives to generate functional nanotubes at room temperature.

A controlled functionalization of multi walled carbon nanotubes through Diels–Alder cycloaddition reaction of various substituted benzocyclobutenes was reported by Bhaskaran et al [127]. Both bulk and solution functionalization conditions along with different functionalizing temperature were used to investigate the extent of control over the modification process. The stained ring of the benzocyclobutene is susceptible to thermal ring opening to form o-quinodimethane which can undergo Diels–Alder cycloaddition readily. Various substituted benzocyclobutenes were reacted with multi walled carbon nanotubes to quantify the modification mechanisms. It was reported that the extent of functionalization could be controlled by using the appropriate temperature and mode of addition of the reagent and by controlling the reaction times. Additionally, it was observed that the use of polymer containing pendant benzocyclobutene produced cross-linked organic nanoclusters covalently connected to the surface of the nanotubes.

## References

Ajayan, PM; Tour, JM. *Nature*, 2007, 447, 1066.

In *Carbon Nanotubes: Synthesis, Structure Properties, and Applications* ; Dresselhaus, M. S; Dresselhaus, G; Avouris, P; Eds; Springer-Verlag: Berlin, Germany, 2001.

Baughman, RH; Zakhidov, AA; Heer, WA. *Science*, 2002, 297, 787.

Avouris, P. *Acc. Chem. Res*, 2002, 35, 1026.

Iijima, S. *Nature*, 1991, 354, 56.

Endo, M. *PhD Thesis*, Nagoya University, Japan, 1978.

Harris, PJF. *Carbon nanotubes and related structures new materials for the twenty-first century*; Cambridge University Press: Cambridge, 2001.

O'Connell, MJ. *Carbon nanotubes properties and applications*; CRC Taylor & Francis: Boca Raton, 2006.

Harris, PJF. *Carbon nanotubes and related structures*; Cambridge University Press: Cambridge, 2001.

Journet, C; Master, WK; Bernier, P; Loiseau, A; Lamy de la Chapelle, M; Lefrant, S; Deniard, P; Lee, R; Fischer, J. E. *Nature,* 1997, 388, 756.

Ebbesen, TW; Ajayan, PM. *Nature,* 1992, 358, 220.

Colbert, DT; Zhang, J; McClure, SM; Nikolaev, P; Chen, Z; Hafner, JH; Owens, DW; Kotula, PG; Carter, CB; Weaver, JH; Rinzler, AG; Smalley, RE. *Science,* 1994, 266, 1218.

Thess, A; Lee, R; Nikolaev, P; Dai, H; Petit, P; Robert, J; Xu, C; Lee, YH; Kim, SG; Rinzler, AG; Colbert, DT; Scuseria, GE; Tomanek, D; Fischer, JE; Smalley, RE. *Science*, 1996, 273, 483.

Cheng, HM; Li, F; Su, G; Pan, HY; He, LL; Sun, X; Dresselhaus, MS. *Appl. Phys. Lett*, 1998, 72, 3282.

Rao, CNR; Govindaraj, A; Sen, R; Satishkumar, BC. *Mater. Res. Innov*, 1998, 2, 128.

Andrews, R; Jacques, D; Rao, AM; Derbyshire, F; Qian, D; Fan, X; Dickey, EC; Chen, J. *Chem. Phys. Letters,* 1999, 303, 467.

Ren, ZF; Huang, ZP; Xu, J. W; Wang, JH; Bush, P; Siegel, MP; Provenico, PN. *Science,* 1998, 282, 1105.

Singh, C; Shaffer, MSP; Kinloch, IA; Windle, AH. *Physica B,* 2002, 323, 339.

Lau, KT; Hui, D. *Comp. Part B: Eng.,* 2002, 33, 263.

Li, F; Cheng, HM; Bai, S; Su, G; Dresselhaus, MS. *Appl. Phys. Lett.,* 2000, 77, 3161.

Dresselhaus, MS; Dresselhaus, G; Eklund, PC. *Science of Fullerenes and Carbon Nanotubes*; Academic Press: New York, 1996.

Treacy, MM. J; Ebbesen, TW; Gibson, JM. *Nature,* 1996, 381, 678.

Wong, EW; Sheehan, PE; Lieber, CM. *Science,* 1997, 277, 1971.

Poncharal, P; Wang, ZL; Ugarte, D; de Heer, WA. *Science,* 1999, 283, 1513.

Hone, J; Whitney, M; Piskoti, C; Zettl, A. *Phys. Rev., B* 1999, 59, R2514.

O´Connell, MJ; Boul, P; Ericson, LM; Huffman, C; Wang, Y; Haroz, E; Kuper, C; Tour, J; Ausman, KD; Smalley, RE. *Chem. Phys. Lett.,* 2001, 342, 265.

Vedala, H; Choi, YC; Kim, G; Lee, E; Choi, WB. *Proceedings of 2nd International Latin American and Caribbean Conference for Engineering and Technology*; Florida, USA.

Star, A; Stoddart, JF; Steuerman, D; Diehl, M; Boukai, A; Wong, EW; Yang, X; Chung, SW; Choi, H; Heath, JR. *Angew. Chem. Int. Ed.,* 2001, 40, 1721.

Star, A; Liu, Y; Grant, K; Ridvan, L; Soddart, JF; Steuerman, DW; Diehl, MR; Boukai, A; Heath, JR. *Macromolecules*, 2003, 36, 553.

Zhu, J; Yudasaka, M; Zhang, M; Ijima, S. *J. Phys. Chem B,* 2004, 108, 11317.

Star, A; Steurmann, DW; Heath, JR; Stoddart, JF. *Angew. Chem. Int. Ed*, 2002, 41, 2508.

Bandyopadhyaya, R; Nativ-Roth, E; Regev, O; Yerushalmi-Rozen, R. *Nano Lett*, 2002, 2, 25.

Shim, M; Kam, NWS; Chen, RJ; Li, Y; Dai, H. *Nano Lett*, 2002, 2, 285.

Shim, M; Javey, A; Kam, NWS; Dai, H. *J. Am. Chem. Soc*, 2001, 123, 11512.

Harris, JM; Zalipsky, S. *Poly(ethylene glycol): Chemistry and Biological Application*; American Chemical Society: Washington DC, 1997.

In *Biocompatible polymers, metals and composites*; Szycher, M; Ed; Technomic: Lancaster PA, 1983.

Ostuni, E; Chapman, RG; Holmlin, RE; Takayama, S; Whitesides, GM. *Langmuir* 2001, 17, 5605.

Fu, Q; Lu, C; Liu, J. *Nano Lett*, 2002, 2, 329.

Seeger, T; Köhler, T; Frauenheim, T. ; Grobert, N; Rühle, M; Terrones, M; Seifert, G. *Chem. Commun*, 2002, 34.

Gomez, FJ; Chen, RJ; Wang, D; Waymouth, RM; Dai, H. *Chem. Commun*, 2003, 190.

Carillo, A; Swartz, JA; Gambe, JM; Kane, RS; Chakrapani, N; Wei, B; Ajayan, PM. *Nano Lett*, 2003, 3, 1437.

Decher, G; Lvov, Y; Schmitt, J. *Thin Solid Films*, 1994, 244, 772.

Stroock, AD; Kane, RS; Weck, M; Metallo, SJ; Whitesides, GM. *Langmuir*, 2003, 10, 2466.

Korneev, D; Lvov, Y; Decher, G; Schmitt, J; Yaradaikin, S. *Physica B*, 2005, 213-214, 954.

Decher, G. *Science*, 1997, 277, 1232.

Hammond, P. T. *Curr. Opin. Colloid Interface Sci*, 1999, 4, 430.

Schmitt, J; Grünewald, T; Decher, G; Pershan, PS; Kjaer, K; Lösche, M. *Macromolecules*, 1993, 26, 7058.

Lahiri, J; Isaacs, L; Tien, J; Whitesides, G. M. *Anal. Chem*, 1999, 71, 777.

Wang, TC; Chen, B; Rubner, MF; Cohen, RE. *Langmuir*, 2001, 17, 6610.

Caruso, F; Caruso, RA; Mohwald, H. *Science*, 1998, 282, 1111.

Chapman, RG; Ostuni, E; Liang, MN; Meluleni, G; Kim, E; Yan, L; Pier, G; Warren, HS; Whitesides, GM. *Langmuir*, 2001, 17, 1225.

Kang, Y; Taton, T. A. *J. Am. Chem, Soc.*, 2003, 125, 5650.

Huang, HY; Kowalewski, T; Remsen EE; Gertzmann, R; Wooley, KL. *J. Am. Chem. Soc*, 1997, 119, 11653.

Zhang, L; Shen, H; Eisenberg, A. *Macromolecules*, 1997, 30, 1001.

Yang, Z; Chen, X; Chen, C; Li, W; Zhang, H; Xu, L; Yi, B. *Polym. Compos*, 2007, doi: 10.1002/pc.20254.

Chen, RJ; Zhang, Y; Wang D; Dai, H. *J. Am. Chem. Soc*, 2001, 123, 3838.

Katz, E. *J. Electroanal. Chem*, 1994, 365, 157.

Jaegfeldt, H; Kuwana, T; Johansson, G. *J. Am. Chem. Soc*, 1983, 105, 1805.

Liu, YT; Zhao, W; Huang, ZY; Gao, YF; Xie, XM; Wang, XH; Ye, XY. *Carbon*, 2006, 1581.

Liu, Y; Wu, DC; Zhang, YD; Jiang, X; He, CB; Chung, TS; Goh SH; Leong KW. *Angew. Chem. Int. Ed*, 2005, 44, 4782.

Wang, Y; Iqbal, Z; Malhotra, SV. *Chem. Phys. Lett*, 2005, 402, 96.

Guiseppi-Elie, A; Lei, C; Baughman, RH. *Nanotechnol.*, 2002, 13, 559.

Wang, SG; Zhang, Q; Wang, R; Yoon, SF; Ahn, J; Yang, DJ; Tian, JZ; Li, J. Q; Zhou, Q. *Electrochem. Commun*, 2003, 5, 800.

Wang, SG; Zhang, Q; Wang, R; Yoon, SF. *Biochem. Biophys. Res. Commun*, 2003, 311, 572.

Yu, X; Chattopadhyay, D; Galeska, I; Papadimitrakopoulos, F; Rusling, JF. *Electrochem. Commun*, 2003, 5, 408.

Huang, W; Fernando, S; Allard, LF; Sin, YP. *Nano Lett*, 2003, 3, 565.

Velasco-Santos, C; Martinez-Hernandez, AL; Fischer, FT; Ruoff, R; Castano, VM. *Chem. Mater*, 2003, 15, 4470.

Jia, Z; Wang, Z; Xu, C; Liang, J; Wei, B; Wu, D; Zhu, S. *Mater. Sci. Eng*, 1999, A271, 395.

Geng, HZ; Rosen, R; Zheng, B; Shimoda, H; Fleming, L; Liu, J; Zhou, O. *Adv. Mater.*, 2002, 14, 1387.

Qin, S; Qin, D; Ford, WT; Resasco, DE; Herrera, JE. *J. Am. Chem. Soc.*, 2004, 126, 170.

Zhao, B; Brittain, WJ. *Prog. Polym. Sci*, 2000, 25, 677.

Huang, W; Skanth, G; Baker, GL; Bruening, ML. *Langmuir*, 2001, 17, 1731.

Husseman, M; Malmstroem, EE; McNamara, M; Mate, M; Mecerreyes, D; Benoit, D. G; Hedrick, JL; Mansky, P; Huang, E; Russell, TP; Hawker, CJ. *Macromolecules* 1999, 32, 1424.

Ejaz, M; Yamamoto, S; Ohno, K; Tsujii, Y; Fukuda, T. *Macromolecules*, 1998, 31, 5934.

Boerner, HG; Beers, K; Matyjaszewski, K; Sheiko, SS; Moeller, M. *Macromolecules* 2001, 34, 4375.

Milner, ST. *Science*, 1991, 251, 905.

Matyjaszewski, K; Miller, PJ; Shukla, N; Immaraporn, B; Gelman, A; Luokala, BB; Siclovan, TM; Kickelbick, G; Vallant, T; Hoffmann, H; Pakula, T. *Macromolecules*, 1999, 32, 8716.

Vestal, CR; Zhang, ZJJ. *J. Am. Chem. Soc*, 2002, 124, 14312.

Jordan, R; Ulman, A. *J. Am. Chem. Soc*, 1998, 120, 243.

Jordan, R; Ulman, A; Kang, JF; Rafailovich, MH; Sokolov, J. *J. Am. Chem. Soc.*, 1999, 121, 1016.

Weck, M; Jackiw, JJ; Rossi, RR; Weiss, PS; Grubbs, RH. *J. Am. Chem. Soc*, 1999, 121, 4088.

Mittal, V. *J. Colloid Interface Sci*, 2007, 314, 141.

Kong, H; Gao, C; Yan, D. *J. Am. Chem. Soc.* 2004, 126, 412.

Shi, D; Lian, J; He, P; Wang, LM; van Ooij, WJ; Schulz, M; Liu, Y; Mast, DB. *Appl. Phys. Lett.*, 2002, 81, 5216.

Eitan, A; Jiang, K; Dukes, D; Andrews, R; Schadler, LS. *Chem. Mater*, 2003, 15, 3198.

Eitan, A; Fisher, FT; Andrews, R; Brinson, LC; Schadler, LS. *Comp. Sci. Technol*, 2006, 66, 1159.

Li, MS; Ma, CC. M; Lin, ML; Chang, FC. *Polymer*, 1997, 38, 4903.

Su, CC; Woo, EM; Chen, CY; Wu, RR. *Polymer*, 1997, 38, 2047.

Ding, W; Eitan, A; Fisher, FT; Chen, X; Dikin, DA; Andrews, R; Brinson, LC; Schadler, LS; Ruoff, RS. *Nano Lett*, 2003, 3, 1593.

Mylvaganam, K; Zhang, LC. *J. Phys. Chem. B*, 2004, 108, 15009.

Viswanathan, G; Chakrapani, N; Yang, H; Wei, B; Chung, H; Cho, K; Ryu, CY; Ajayan, PM. *J. Am. Chem, Soc.*, 2003, 125, 9258.

Bauschlicher, CW. *Chem. Phys. Lett*, 2000, 322, 237.

Bauschlicher, CW. *Nano Lett*, 2001, 1, 223.

Jaffe, RL. *J. Phys. Chem B*, 2003, 107, 10378.

Mylvaganam, K; Zhang, LC. *J. Phys. Chem. B*, 2004, 108, 5217.

Vosko, SH; Wilk, L; Nusair, M. *Can. J. Phys.*, 1980, 58, 1200.

Becke, AD. *J. Chem. Phys.*, 1993, 98, 5648.

Mulliken, RS. *J. Chem. Phys.*, 1955, 23, 1833.

Liu, YX; Du, ZJ; Li, Y; Zhang, C; Li, CJ; Yang, XP; Li, HQ. *J. Polym. Sci., Part A: Polym. Chem*, 2006, 44, 6880.

Pan, B; Cui, D; He, R; Gao, F; Zhang, Y. *Chem. Phys. Lett*, 2006, 417, 419.

Sapsford, KE; Medintz, IL; Golden, JP; Deschamps, JR; Uyeda, HT; Mattoussi, H. *Langmuir*, 2004, 20, 7720.

Bachand, GD; Rivera, SB; Boal, AK; Gaudioso, J; Liu, Y; Bunker, BC. *Nano Lett*, 2004, 4, 817.

Wojdel, JC., Bromley, ST. *J. Phys. Chem. B*, 2005, 109, 1387.

Chan, WCW; Nie, S. *Science*, 1998, 281, 2016.

Ge, JJ., Zhang, D; Li, Q; Hou, H; Graham, MJ; Dai, L; Harris, FW; Cheng, SZ. D. *J. Am. Chem. Soc*, 2005, 127, 9984.

Viswanathan, G; Chakrapani, N; Yang, H; Wei, B; Chung, H; Cho, K; Ryu, CY; Ajayan, PM. *J. Am. Chem. Soc*, 2003, 125, 9258.

Khabashesku, VN; Billups, WE; Margrave, JL. *Acc. Chem. Res*, 2002, 35, 1087.
Mickelson, ET; Huffman, CB; Rinzler, AG; Smalley, RE; Hauge, RH; Margrave, JL. *Chem. Phys. Lett*, 1998, 296, 188.
Zhu, J; Kim, JD; Peng, H; Margrave, JL; Khabashesku, VN; Barrera, EV. *Nano Lett.* 2003, 3, 1107.
Khabashesku, VN; Billups, WE; Margrave, JL. *Acc. Chem. Res*, 2002, 35, 1087.
Khabashesku, VN; Margrave, JL. In *Encyclopedia of Nanoscience and Nanotechnology*; Nalwa, HS; Ed; American Scientific Publishers: Stevenson Ranch CA, 2004.
Pulikkathara, MX; Kuznetsov, OV; Khabashesku, VN. *Chem. Mater*, 2008, 20, 2685.
Kakade, BA; Pillai, VK. *Appl. Surf. Sci.* 2008, 254, 4936.
Pan, B; Gao, F; Gu, H. *J. Colloid Interface Sci*, 2005, 284, 1.
Yoza, B; Arakaki, A; Matsunaga, T. *J. Biotechnol*, 2003, 101, 219.
Xu, Y; Gao, C; Kong, H; Yan, D; Jin, YZ; Watts, PCP. *Macromolecules*, 2004, 37, 8846.
Xu, P; Cui, D; Pan, B; Gao, F; He, R; Li, Q; Huang, T; Bao, C; Yang, H. *Appl. Surf. Sci.*, 2008, 254, 5236.
Ke, G; Guan, W; Tang, C; Guan, W; Zeng, D; Deng, F. *Biomacromolecules*, 2007, 8, 322.
Tseng, CH; Wang, CC; Chen, CY. *Chem. Mater*, 2007, 19, 308.
Yang, Z; Chen, XH; Xia, SZ; Pu, YX; Xu. HY; Li, WH; Xu, LS; Yi, B; Pan, WY. *J. Mater. Sci*, 2007, 42, 9447.
Yang, Z; Chen, XH; Liu, YQ; Chen, XH; Chen, CS; Li, WH; Xu, LS. *Acta Chin. Sinca*, 2006, 64, 203.
Kong, H; Luo, P; Gao, C; Yan, D. *Polymer*, 2006, 46, 2472.
Kong, H; Gao, C; Yan, D. *J. Mater. Chem.*, 2004, 14, 1401.
Chen, SM; Wu, GZ; Liu, YD; Long, DW. *Macromolecules*, 2006, 39, 330.
Wei, L; Zhang, Y. *Chem. Phys, Lett.*, 2007, 446, 142.
Martinez-Rubi, Y; Guan, J; Lin, S; Scriver, C; Sturgeon, RE; Simard, B. *Chem. Commun*, 2007, 5146.
Sakellariou, G; Ji, H; Mays, JW; Hadjichristidis, N; Baskaran, D. *Chem. Mater.*, 2007, 19, 6370.

In: Polymer Nanocomposites: Advances in Filler Surface
Editor: Vikas Mittal
ISBN: 978-1-60876-125-8


*Chapter X*

# Exchange of Chain-end Functionalized Polyolefins on the Clay Surface

***T. C. Mike Chung****

Department of Materials Science and Engineering
The Pennsylvania State University University Park, PA, USA

## Abstract

This chapter discusses a unique chain-end functionalized polyolefin structure that contains a reactive terminal group for forming a strong interaction with clay surfaces, resulting in exfoliated polyolefin/clay nanocomposites. Two commercially important polymers, including isotactic polypropylene (PP) and poly(vinylidene fluoride) (PVDF), have been studied to illustrate the research idea and experimental results. The chemistry to prepare chain-end functionalized PP containing a terminal OH or $NH_3^+Cl^-$ group involves metallocene-mediated propylene polymerization using rac-$Me_2$Si[2-Me-4-Ph(Ind)]$_2$Zr$Cl_2$/MAO complex in the presence of styrene derivatives (St-f) and hydrogen, which serve as the chain transfer agents. On the other hand, the chain end functionalized fluoropolymer is prepared by a specific borane control radical initiator that bears a reactive $Si(OR)_3$ group that is directly transferred to the beginning of the polymer chain. Despite the extremely low concentration of the chain-end group in PP and PVDF polymers, the terminal OH, $NH_3^+Cl^-$, and $Si(OR)_3$ groups show good accessibility and reactivity toward the clay surfaces and form hydrogen bond, ion-exchange the cations ($Li^+$, $Na^+$, etc.), and chemical bonding, respectively. Such strong interactions anchor the PP or PVDF chain to the clay surfaces. On the other hand, the remaining unperturbed end-tethered high molecular weight PP or PVDF tail exfoliates the clay layers. This exfoliated structure is maintained even after further mixing with pure neat PP or PVDF polymers.

**Keywords:** functionalization; polypropylene; poly(vinylidene fluoride); metallocene; control radical polymerization, nanocomposite; PP/clay composite, PVDF/clay nanocomposite.

# 10.1. Introduction

Polyolefin thermoplastics, such as isotactic polypropylene (PP), poly(vinylidene fluoride) (PVDF), is a very important class of materials, accounting for almost 50% of the total polymers produced. Polyolefin thermoplastic possesses an excellent combination of mechanical, chemical and electronic properties, along with low cost, light weight (low density) and superior processability and recyclability. Its properties would be further widened if it could be mixed with other materials. Unfortunately, it is also known for being a challenging material in forming uniform polymer blends and composites, due to poor interactive (inert) properties and low surface energy [1]. The combination of lack of polar functionality and high crystallinity-resulting in poor compatibility with other materials-has been a major drawback in many applications, particularly with those in which adhesion or compatibility with other materials is paramount [2]. It is even more difficult to prepare polyolefin/clay nanocomposites with exfoliated structures [3]. In addition to the extreme high interfacial area, it is very difficult to maintain stability of the dispersed hydrophilic clay sheets that are imbedded in an antithetical hydrophobic polyolefin matrix under various processing and application environments. Scientifically, it has been most interesting to discern the most effective interfacial agent and structure that can achieve stable exfoliated polyolefin/clay nanocomposites, especially one that could be applied under conventional processing conditions or existing procedures to achieve enhanced properties.

Based on theoretical models, Balazs [4] suggested that an increase in the length of the organic molecules tethered on clay is a promising approach to further promote dispersion of clay sheets in a polymer matrix. In some reports [5] in which a variety of polymers were employed—such as polystyrene, poly(methyl methacrylate), and poly(ε-caprolactone), grafted on montmorillonite (MMT) surfaces—varied extents of clay dispersion have been observed. Furthermore, subsequent annealing of these polymer-bearing clays, or mixing them with the respective homopolymers, can lead to retention or collapse of the clay dispersion, depending on the polymer. The differences in these responses were attributed to the specific polymer-clay interactions in each case; where there exists strong interactions, the clay dispersions collapse, whereas the dispersions can be retained in the cases of weak polymer-clay interactions. Logically, the chain-end functionalized polyolefin should be a very attractive interfacial agent, possessing an unperturbed polymer chain structure and desirable physical properties (such as melting temperature, crystallinity, glass transition temperature, melt flow, etc.) that are identical to the pure (neat) polymer. Nevertheless, a terminal reactive functional group at the polymer chain-end with good mobility (accessibility) and reactivity can form a strong interaction [6] or chemical bond with inorganic substrates during reactive blending. In other words, the suitable chain-end functionalized polymer should effectively secure two surfaces, with excellent compatibility (co-crystallization) with the corresponding neat polymer matrix and good adhesion on inorganic substrates.

In general, the functionalization of polyolefins has long been a scientifically challenging and technologically important research area. In the past, the preparation of chain-end functionalized polymers has been limited to living polymerization with the use of an initiator bearing a

* Corresponding author: E-mail: chung@ems.psu.edu

functional group [7] or a control termination (functionalization) reaction [8] of the living polymer chain-end. Unfortunately, there are only a few Ziegler-Natta type transition metal coordination catalysts that exhibit living polymerization behavior, and most of them are limited to the preparation of polyethylene and poly(1-hexene) [9]. Furthermore, living polymerization only produces one polymer chain per initiator, which presents a relatively low rate of catalyst activity in the typical polyolefin preparation.

In the past decade, we have been studying new chemical routes [1] to prepare functional polyolefins (including PE, PP, s-PS, PVDF, etc.), including chain-end functionalization chemistry. The incorporation of a terminal functional group into a PE, PP, or s-PS chain is based on a chain transfer reaction involving a reactive chain transfer agent - such as dialkylboranes [10] and styrene derivatives/hydrogen [11] - during metallocene-mediated olefin polymerization. All polymers formed contain a terminal borane or styrene derivative group and have a relatively narrow molecular weight distribution ($M_w/M_n$~2). The polymer molecular weight (from a few thousand to a hundred thousand) was essentially controlled by the mole ratio of [chain transfer agent]/[olefin]. Both the terminal borane and styrene derivative are very versatile, and can serve as the reactive sites for subsequent functionalization reactions or conversion to living initiators for chain extension reactions. A few years ago, we reported on a borane/oxygen radical initiator system [12] that carries out a control radical polymerization of acrylic monomers in bulk and in solution at ambient temperature. The same borane initiators were effective for fluoromonomers, such as vinylidene fluoride (VDF), trifluoroethylene (TrFE), and chlorotrifluoroethylene (CTFE), which prepare the fluoropolymer and random copolymers in mild reaction conditions [13]. Recently, we extended the chemistry to functional borane initiators with the objective of preparing chain-end functional fluoropolymers [14].

## 10.2. Synthesis of PP-t-Cl, PP-t-OH and PP-t-$NH_2$

It is very desirable to extend the metallocene-mediated chain transfer reaction to directly prepare the polyolefin with a desirable terminal functional group, such as Cl, OH, and $NH_2$ [11]. In other words, the ideal reaction is a one-pot *in-situ* polymerization process, and no chain end functionalization would be needed after the polymerization reaction. The basic idea in the direct preparation of the chain functionalized PP was to use a functionalized styrenic chain transfer agent (St-f). The chain transfer agent could engage in a similar metallocene-mediated propylene polymerization/chain transfer reaction under some reaction conditions. Three St-f molecules were investigated, including p-chlorostyrene (St-Cl), silane-protected p-vinylphenol (St-OSi), and silane-protected p-ethylaminostyrene (St-$NSi_2$), as illustrated in Scheme 1. The silane groups provide effective protection for both the OH and $NH_2$ functional groups during the metallocene catalysis, and can be completely de-protected by aqueous HCl solution during the sample work-up procedure.

No external protection agent is needed for St-Cl in metallocene/MAO systems. However, the metallocene cationic site is very sensitive to the OH and $NH_2$ functional groups. The silane groups not only provide effective protection for both the OH and $NH_2$ functional groups during the metallocene catalysis, but also can be completely deprotected by aqueous

HCl solution during the sample work-up procedure. The overall reaction especially benefits from the very small quantity of the chain transfer agent needed in the preparation of high polymers. Therefore, the additional protection-deprotection step causes almost no change in the polymerization conditions.

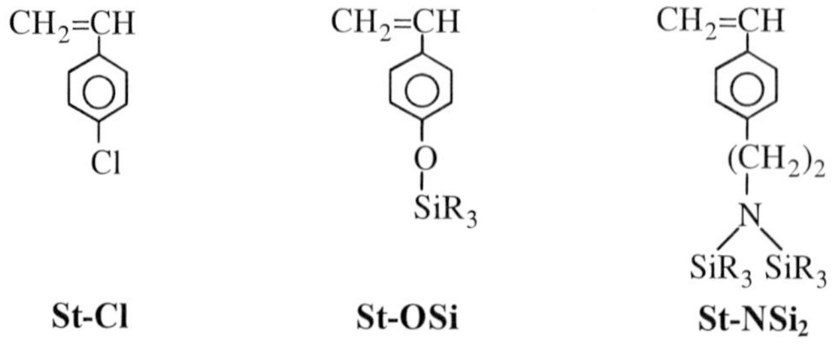


Scheme 1.

Equation 1 illustrates the general reaction scheme. During the course of propylene 1,2-insertion, the propagating $M^+$-C site (II) eventually reacts with a St-NSi$_2$ unit ($k_{12}$) (via 2,1-insertion) to form a St-NSi$_2$ capped propagating site (III) with an adjacent phenyl group interacting with metal cation. The new propagating site (III) is incapable of continuing the insertion of St-NSi$_2$ ($k_{22}$) or propylene ($k_{21}$) due to the steric jamming. However, it can react with hydrogen to complete the chain transfer reaction. This consecutive reaction with St-NSi$_2$ and hydrogen results in a PP-t-St-NSi$_2$ polymer chain (IV) and a regenerated Zr-H species (I) reinitiates the polymerization of propylene and continues the polymerization cycles. After completing the polymerization, the desirable $NH_2$ terminal group in PP-t-$NH_2$ (V) can be easily recovered during the sample work-up step using HCl aqueous solution. Theoretically, the PP-t- $NH_2$ molecular weight should be linearly proportion to the molar ratio of [propylene]/[St-NSi$_2$].

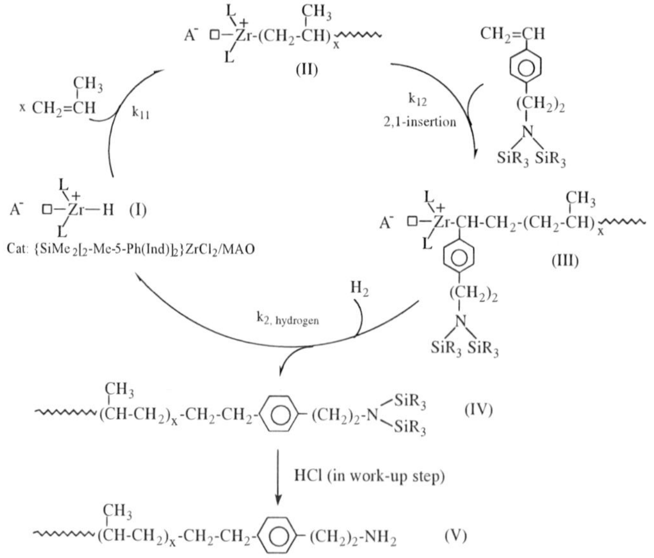


Equation 1.

Table 1 summarizes the experimental results involving St-Cl/$H_2$, St-OSi/$H_2$, and St-$NSi_2$/$H_2$ in the rac-$Me_2Si[2\text{-}Me\text{-}4\text{-}Ph(Ind)]_2ZrCl_2$/MAO catalyzed polymerization of propylene. In all control reactions, a small amount of any styrene derivative (St-f) effectively stops the polymerization of propylene. The introduction of hydrogen gradually restores the catalyst activity. Overall, the PP molecular weight is governed by the chain transfer agent; the higher the concentration of the St-f, the lower the molecular weight of the resulting polymer.

**Table 1. A summary of PP-t-Cl, PP-t-St-OH and PP-t-St-$NH_2$ polymers[a]**

| Run | St-f[b] (mol/l) | $H_2$ (psi) | Yield (g) | Cat. Activity[c] | Cl, OH or $NH_2$ in PP (mol%) | $M_n$ ($\times10^{-3}$) | PDI ($M_w/M_n$) | Tm (°C) |
|---|---|---|---|---|---|---|---|---|
| A-1 | 0.144 | 0 | ~ 0 | ~ 0 | - | - | - | - |
| A-2 | 0.144 | 6 | 1.36 | 4,402 | 0.13 | 45.3 | 1.9 | 159.0 |
| A-3 | 0.144 | 12 | 7.56 | 24,192 | 0.13 | 47.5 | 2.0 | 158.2 |
| A-4 | 0.144 | 20 | 21.8 | 70,435 | 0.12 | 46.1 | 2.1 | 158.1 |
| B-1 | 0.289 | 0 | ~ 0 | ~ 0 | - | - | - | - |
| B-2 | 0.289 | 6 | 1.24 | 4,388 | 0.23 | 18.5 | 2.2 | 156.8 |
| B-3 | 0.289 | 14 | 4.44 | 15,712 | 0.23 | 19.1 | 2.1 | 157.1 |
| B-4 | 0.289 | 20 | 12.91 | 45,673 | 0.22 | 18.7 | 2.1 | 156.9 |
| C-1 | 0.198 | 0 | ~ 0 | ~ 0 | - | - | - | - |
| C-2 | 0.198 | 6 | 0.36 | 1,165 | 0.09 | 52.1 | 2.3 | 159.1 |
| C-3 | 0.198 | 12 | 2.36 | 7,637 | 0.08 | 53.4 | 2.4 | 158.4 |
| C-4 | 0.198 | 20 | 8.34 | 26,995 | 0.10 | 52.5 | 2.2 | 158.2 |
| D-1 | 0.396 | 0 | ~ 0 | ~ 0 | - | - | - | - |
| D-2 | 0.396 | 6 | ~ 0 | ~ 0 | - | - | - | - |
| D-3 | 0.396 | 12 | 0.63 | 2,039 | 0.18 | 23.4 | 2.5 | 157.4 |
| D-4 | 0.396 | 20 | 2.57 | 8,318 | 0.20 | 22.0 | 2.0 | 156.3 |
| E-1 | 0.125 | 0 | ~ 0 | ~ 0 | - | - | - | - |
| E-2 | 0.125 | 6 | 0.41 | 1,327 | 0.07 | 56.3 | 2.1 | 159.1 |
| E-3 | 0.125 | 12 | 2.62 | 8,480 | 0.06 | 55.4 | 2.5 | 158.4 |
| E-4 | 0.125 | 20 | 9.78 | 31,655 | 0.08 | 58.9 | 2.3 | 158.2 |
| F-1 | 0.250 | 0 | ~ 0 | ~ 0 | - | - | - | - |
| F-2 | 0.250 | 12 | 0.81 | 2,622 | 0.18 | 25.3 | 2.2 | 156.3 |
| F-3 | 0.250 | 20 | 3.11 | 10,066 | 0.19 | 24.2 | 2.3 | 155.9 |

a) Reaction conditions: 50 ml toluene, Catalyst: *rac*-$Me_2Si[2\text{-}Me\text{-}4\text{-}Ph(Ind)]_2ZrCl_2$/ MAO catalyst, [Zr]= $1.25\times10^{-6}$ moles, [MAO]/[Zr]= 3000, propylene= 100 psi, temperature= 30°C, time= 15 min.
b) St-f: St-Cl for runs A and B; St-OSi for Runs C and D, St-$NSi_2$ for Runs E and F.
c) Catalyst activity = kg of PP/ mol of catalyst.h.

Figure 1 shows the plot of the polymer molecular weight (Mn) versus the molar ratio of [propylene]/[St-f], including three St-Cl/$H_2$, St-OSi/$H_2$, and St-$NSi_2$/$H_2$ chain transfer agents. In general, the polymer molecular weight and molar ratio of [propylene]/[St-f] are linearly proportional. It is clear that the chain transfer reaction to St-f (with rate constant $k_{tr}$) is the dominant termination process, and that it competes with the propagating reaction (with the rate constant $k_p$). The degree of polymerization ($X_n$) follows a simple comparative equation

$X_n=k_p[olefin]/k_{tr}[p\text{-}MS]$ with a chain transfer constant $k_{tr}/k_p$ of 1/21 for St-Cl/$H_2$, 1/48 for St-OSi/$H_2$, and 1/34 for St-$NSi_2$/$H_2$. It is intriguing that the $k_{tr}/k_p$ values are significantly lower than those seen in styrene and p-methylstyrene cases [11] under similar reaction conditions. The bulky protected functional groups may reduce the frequency of the chain transfer reaction. Polymers with very low molecular weights (a few thousand) have been obtained, and the molecular weight distributions are quite narrow, which is generally consistent with single site polymerization processes.

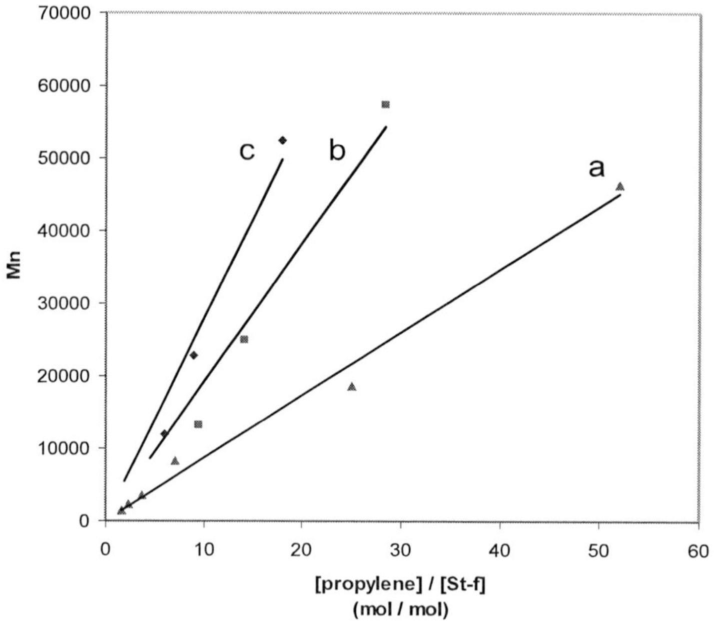


Figure 1. The plots of number average molecular weights (Mn) of PE-t-St-f polymers versus the mole ratio of [propylene]/[St-f] using (a) St-Cl, (b) St-OSi and (c) St-$NSi_2$, respectively. [Reproduced with permission from Macromolecules 2002, 35, 9352-9359. Copyright 2002 Am. Chem. Soc.].

The terminal functional group at the polymer chain end provides direct evidence for the chain transfer reaction. Figure 2 shows the $^1$H NMR spectra (with inset of magnified region and chemical shift assigments) of PP-t-St-OSi polymer and the corresponding PP-t-St-OH. In addition to three major peaks ($\delta$= 0.95, 1.35, and 1.65 ppm) for the $\mathbf{CH_3}$, $\mathbf{CH_2}$, and $\mathbf{CH}$ groups in the PP backbone, there are three minor chemical shifts at 0.25, 2.61, and 6.75-7.18 ppm (with an intensity ratio near 6/2/4) shown in Figure 2(a), corresponding to -OSi($\mathbf{CH_3}$)$_2$(t-Bu), –$\mathbf{CH_2}$-$\phi$, and -$C_6\mathbf{H_4}$-OSi, respectively. The chemical shift for the silane protecting group completely disappears in Figure 2(b), indicating the occurrence of a very effective deprotection reaction. The equally split chemical shifts for the phenyl protons, combined with no detectable side product, further indicate the terminal p-alkylphenol moiety.

Figure 3 compares the $^1$H NMR spectra of PP-t-St-$NSi_2$ polymer (Mn= $24.2\times10^3$; Mw/Mn = 2.3) and the corresponding PP-t-St-$NH_2$. In addition to the chemical shifts for PP polymer, Figure 3(a) shows all of the chemical shifts associated with the protecting bis(trimethylsily)amino terminal group that is connected to the symmetrical p-dialkylbenzene moiety. In fact, all four phenyl protons merge into a single chemical shift at 7.22 ppm. Figure 3(b) shows an almost identical spectrum, except for the disappearance of the silane protecting group at 0.24 ppm.

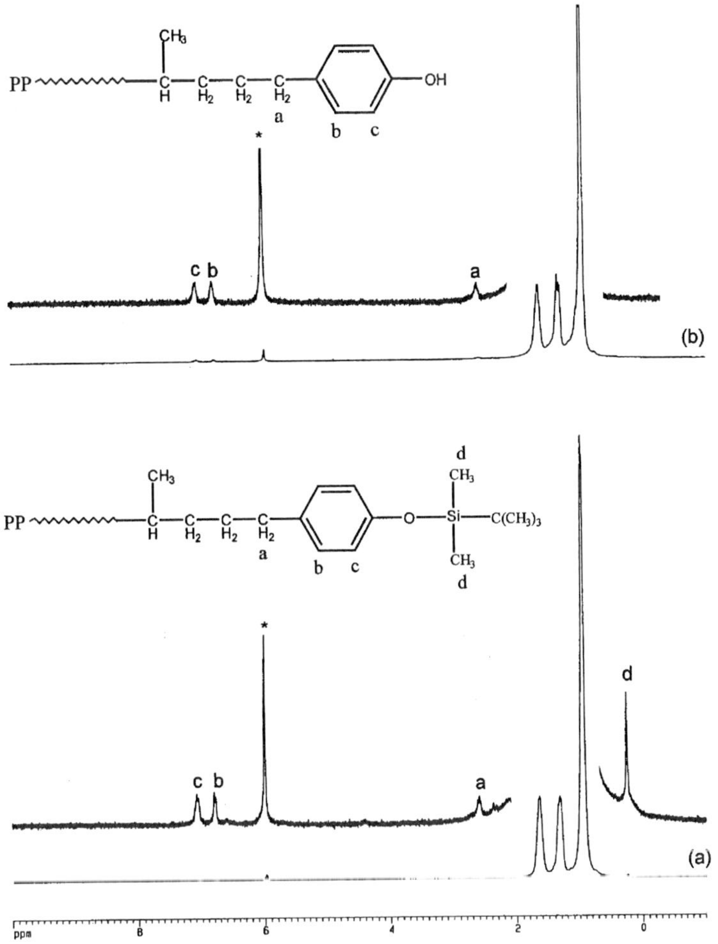


Figure 2. [1]H NMR spectra of (a) a PP-t-St-OSi polymer and its corresponding PP-t-St-OH. (Mn= 22,000 g/mol, Mw/Mn= 2.0). (solvent: $C_2D_2Cl_4$; temp.: 110oC). [Reproduced with permission from Macromolecules 2002, 35, 9352-9359. Copyright 2002 Am. Chem. Soc.].

The existence of a terminal functional group in PP was further evidenced by a chain extension reaction using the terminal functional group as the reaction site for the coupling reaction with polyester and polyamide. Overall, the combined experimental results strongly indicate a clean and effective reaction scheme, as illustrated in Equation 1. The combination of the facile *in situ* chain transfer to St-f/$H_2$-during the catalytic polymerization of propylene and the subsequent complete deprotection reaction during the sample work-up step-begets a highly interesting reaction scheme in the preparation of the chain-end functionalized i-PP with a Cl, OH, and $NH_2$ terminal group via a one-pot reaction process.

# 10.3. Synthesis of PVDF-t-Si(OR)$_3$

For the preparation of chain-end functionalized fluorinated polyolefins [15], we have developed a new class of borane initiators that contain functional groups, such as OH and silane groups, which are stable during the borane oxidation and polymerization processes at ambient temperature. Equation 2 illustrates an example in which the silane-terminated poly(vinylidene fluoride) is prepared with the use of a silane-containing borane initiator at ambient temperature. Overall, this method offers a convenient and effective route to prepare a broad range of chain-end functionalized fluoropolymers under mild reaction conditions.

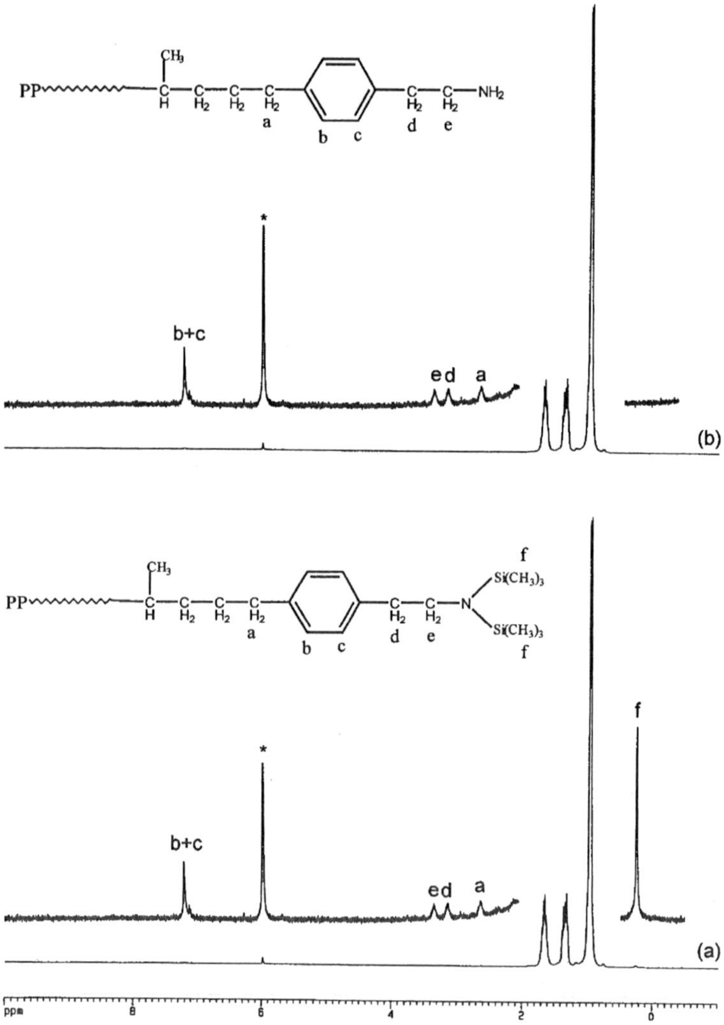


Figure 3. $^1$H NMR spectra of (a) a PP-t-St-NSi$_2$ polymer and (b) its corresponding PP-t-St- NH$_2$ (Mn= 24,200 g/mol, Mw/Mn= 2.3). (solvent: $C_2D_2Cl_4$; temp.: 110$^\circ$ C). [Reproduced with permission from Macromolecules 2002, 35, 9352-9359. Copyright 2002 Am. Chem. Soc.].

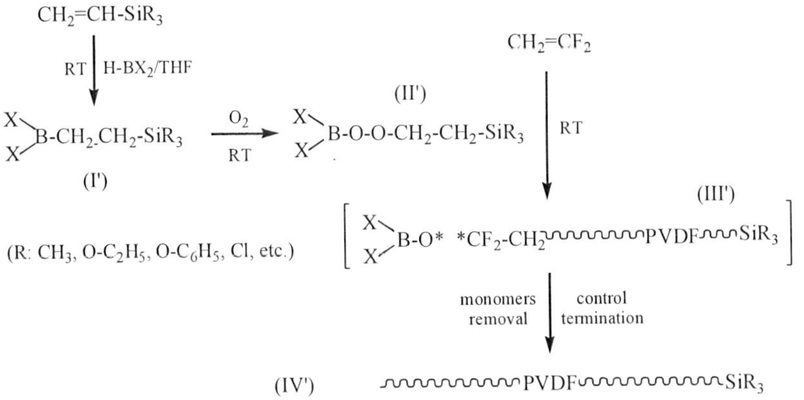


Equation 2

The functional borane initiator (I'), containing a silane group, was quantitatively prepared by a well-known hydroboration reaction [16] of vinylsilane (commercial available) with a borane compound containing at least one B-H group at ambient temperature. The subsequent mono-oxidation reaction of the functional borane initiator, with a control quantity of oxygen, occurs spontaneously at room temperature to form the corresponding peroxylborane (II') for initiating polymerization [12]. This oxidation reaction can be carried out *in situ* during the control radical polymerization with the presence of VDF monomers. The propagating PVDF chain (III') contains a terminal silane group at the beginning of polymer chain and a propagating radical at the other chain-end. The propagating radical at the end of the polymer chain is capped with a borinate radical (*O-$BX_2$) in order to prolong the propagation reaction [12]. Upon releasing VDF monomer gas, the propagating radical is terminated to obtain the silane-terminated PVDF polymer (IV').

Table 2 summarizes the experimental conditions and the results of PVDF polymers prepared by various borane initiators. The first set (runs G-1 to G-5) compares VDF homopolymerization reactions using tri-(triethoxylethylsilyl)borane under various reaction conditions. In comparing runs G-1, G-2, and G-3, the reaction temperature seems to have a relatively small effect to the polymer yield. Both the instantaneous oxygen auto-oxidation of B-C moieties in the borane initiator and the subsequent radical addition reaction with the fluoromonomer are very effective at ambient temperature, even at $0^{o}C$ (run G-1). In comparing runs G-2 and G-4, note that the polymer yield and molecular weight are positively correlated with the monomer conversion. The resulting PVDF-t-$Si(OC_2H_5)_3$ polymer is sensitive to moisture and becomes insoluble upon heating. In the second set (runs H-1 to H-3) we see the preparation of trimethylsilane-terminated PVDF polymers (PVDF-t-$Si(CH_3)_3$) using the tri-(trimethylethylsilyl)borane initiator, with various amounts of the $O_2$ activator ([borane]/[$O_2$]= 2/1 –1/1) at ambient temperature. The resulting PVDF-t-$Si(CH_3)_3$ polymers are moisture-stable and suitable for structural analyses, especially in analyzing end groups and molecular weight. The chain-end structures were determined by the $^{1}H$ NMR spectrum, and the estimated polymer molecular weight from the silane end group analysis was found to be consistent with the GPC result with the help of a universal calibration curve. As will be discussed later, the combined results indicate that most polymer chains contain one terminal silane group.

The functional initiators behave similarly with their corresponding trialkylborane initiators [12] in the polymerization steps. The relatively stable propagating radicals slowly increase polymer molecular weight with monomer conversion, and maintain a relatively narrow molecular weight distribution. The slurry reaction condition, due to the insolubility of PVDF (high crystallinity and high melting temperature) in acetonitrile, may also contribute to the slowdown of the polymerization rate. Note that the relatively slow propagating rate, compared to that of the regular free radical polymerization mechanism, removes the safety concern for heat transport and temperature control usually associated with bulk or solution polymerization of fluoromonomers.

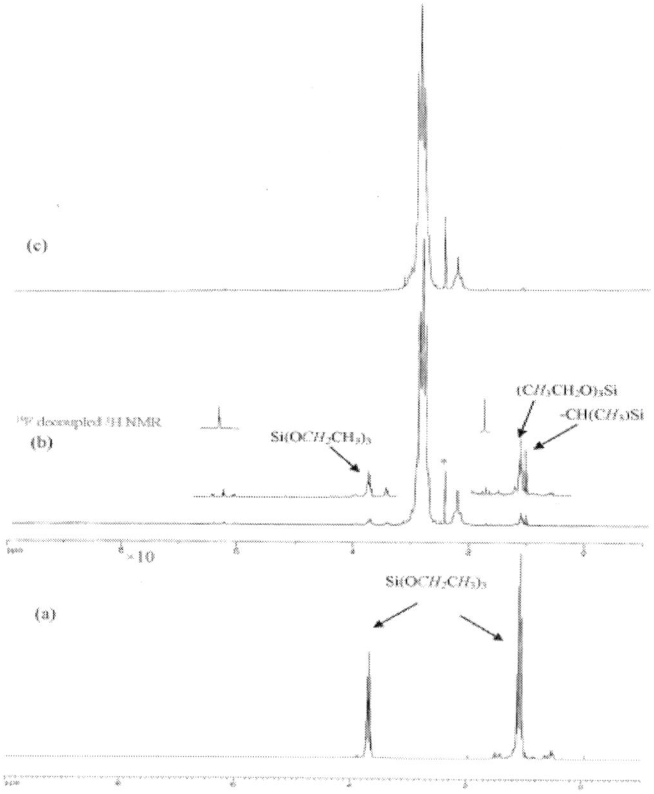


Figure 4. $^{1}H$ NMR spectra of (a) $[(C_2H_5O)_3SiCH_2CH_2]_3B$ initiator, (b) PVDF-t-$Si(OC_2H_5)_3$ polymer (run G-3), and (c) its corresponding hydrolyzed PVDF-t-$Si(OH)_3$ polymer. [Reproduced with permission from Macromolecules 2007, 40, 5235-5240. Copyright 2007 Am. Chem. Soc.].

Figure 4 compares the $^{1}H$ NMR spectra of a PVDF-t-$Si(OC_2H_5)_3$ polymer (run G-3) with the corresponding functional borane initiator containing triethoxyl groups and its hydrolyzed PVDF-t-$Si(OH)_3$ polymer. Insets of the expanded regions exhibit chain end structures and a selected $^{19}F$ decoupled chemical shift. In Figure 4 (b), the PVDF-t-$Si(OC_2H_5)_3$ polymer exhibits two chemical shifts at 2.9 and 2.3 ppm, corresponding to the $CH_2$ units with head-to-tail and head-to-head sequences in the PVDF backbone, respectively. In the expanded regions, there are two dominate chemical shifts at 1.15 ppm ($CH_3$) and 3.75 ppm ($OCH_2$), which are almost identical with those in Figure 4 (a) of the $[(C_2H_5O)_3SiCH_2CH_2]_3B$ initiator.

**Table 2. A summary of chain end functionalized PVDF polymers prepared by functional borane initiators.[a] [Reproduced with permission from Macromolecules 2007, 40, 5235-5240. Copyright 2007 Am. Chem. Soc.].**

| Run | Initiator (mmol) | $O_2$ (mL) | Temp (°C) | Time (hrs) | Conv (%) | End groups (NMR) | | | Mn[b] (NMR) (g/mol) | Mn[c] (NMR) (g/mol) | GPC | |
|---|---|---|---|---|---|---|---|---|---|---|---|---|
| | | | | | | $CF_2H$ (%) | $CH_3$ (%) | $SiR_3$ (%) | | | Mn (g/mol ) | PDI |
| G-1 | B-Si(OEt)$_3$ (1.0) | 25 | 0 | 16 | 32.1 | 32.5 | 9.5 | 58.0 | 22,100 | 38,100 | N.A | N.A |
| G-2 | B-Si(OEt)$_3$ (1.0) | 25 | 25 | 5 | 30.6 | 35.1 | 13.0 | 51.9 | 15,500 | 29,800 | N.A | N.A |
| G-3 | B-Si(OEt)$_3$ (1.0) | 25 | 60 | 5 | 39.6 | 40.2 | 10.1 | 49.7 | 12,100 | 24,700 | N.A | N.A |
| G-4 | B-Si(OEt)$_3$ (1.0) | 25 | 25 | 21 | 51.5 | 33.2 | 13.3 | 54.5 | 22,900 | 42,400 | N.A | N.A |
| G-5 | B-Si(OEt)$_3$ (3.0) | 60 | 25 | 21 | 65.7 | 27.6 | 9.8 | 62.6 | 13,300 | 21,500 | N.A | N.A |
| | | | | | | | | | | | | |
| H-1 | B-SiMe$_3$ (3.0) | 40 | 25 | 5 | 10.1 | 26.4 | 5.3 | 68.3 | 12,500 | 18,400 | 17,500 | 1.4 |
| H-2 | B-SiMe$_3$ (3.0) | 60 | 25 | 5 | 11.2 | 19.3 | 7.9 | 72.8 | 10,500 | 14,600 | 14,500 | 1.6 |
| H-3 | B-SiMe$_3$ (3.0) | 80 | 25 | 5 | 10.2 | 25.4 | 6.3 | 68.3 | 12,900 | 19,000 | 18,200 | 1.4 |
| | | | | | | | | | | | | |
| I-1 | B-Si(OEt)Me$_2$ (1.0) | 25 | 25 | 5 | 30.8 | 38.3 | 11.3 | 50.4 | 18,500 | 37,000 | N.A | N.A |
| I-2 | B-Si(OEt)Me$_2$ (1.0) | 25 | 25 | 17 | 53.3 | 34.4 | 7.0 | 58.6 | 30,000 | 51,700 | N.A | N.A |
| I-3 | B-Si(OEt)$_2$Me (2.0) | 50 | 25 | 17 | 38.8 | 29.8 | 6.8 | 63.4 | 19,600 | 31,100 | N.A | N.A |
| I-4 | B-O-SiMe$_3$ (3.0) | 50 | 25 | 14 | 53.0 | 31.3 | 8.4 | 60.3 | 10,500 | 17,500 | N.A | N.A |
| I-5 | B-SiCl$_3$ (2.0) | 40 | 25 | 19 | 34.3 | N.A | N.A | N.A | N.A | N.A | N.A | N.A |

a. Polymerization conditions: 75mL reactor; 20 ml VDF; solvent: acetonitrile 30mL.
b. Estimated from chain end analysis (including three end groups) by $^1$H NMR spectra.
c. Estimated from cain end analysis (only silane end groups) by $^1$H NMR spectra.

They represent the existence of terminal $(C_2H_5O)_3Si$ silane groups in the PVDF polymer chain end. After hydrolysis by exposing PVDF-t-Si$(OC_2H_5)_3$ polymer in water, the resulting PVDF-t-Si$(OH)_3$ shows almost a complete disappearance of $(C_2H_5O)_3Si$ groups (Figure 4 (c)). Upon drying in vacuum, both PVDF-t-Si$(OEt)_3$ and PVDF-t-Si$(OH)_3$ polymers become insoluble (cross-linked) networks.

One advantage of the borane initiator is its continence in regards to the incorporation of organic functional groups, by the hydroboration reaction of B-H with the α-olefin containing functional group at ambient temperature, as illustrated in Equation 2. In addition, most of the functional groups (and their derivatives) are stable throughout the borane-mediated radical polymerization process. As discussed, the PVDF-t-Si$(OEt)_3$ polymer is sensitive to moisture. The PVDF-t-Si(OEt)$Me_2$ (runs I-1 and I-2) and PVDF-t-Si$(OEt)_2$Me (run I-3) are easier to handle, but the PVDF-t-Si$Cl_3$ (run I-5) is extremely sensitive to moisture, a characteristic that makes it very difficult to maintain good solubility in organic solvents. On the other hand, the PVDF-t-Si$Me_3$ polymers (runs H-1 to H-3) are very stable and are used in examining the polymerization mechanism and the resulting polymer structure.

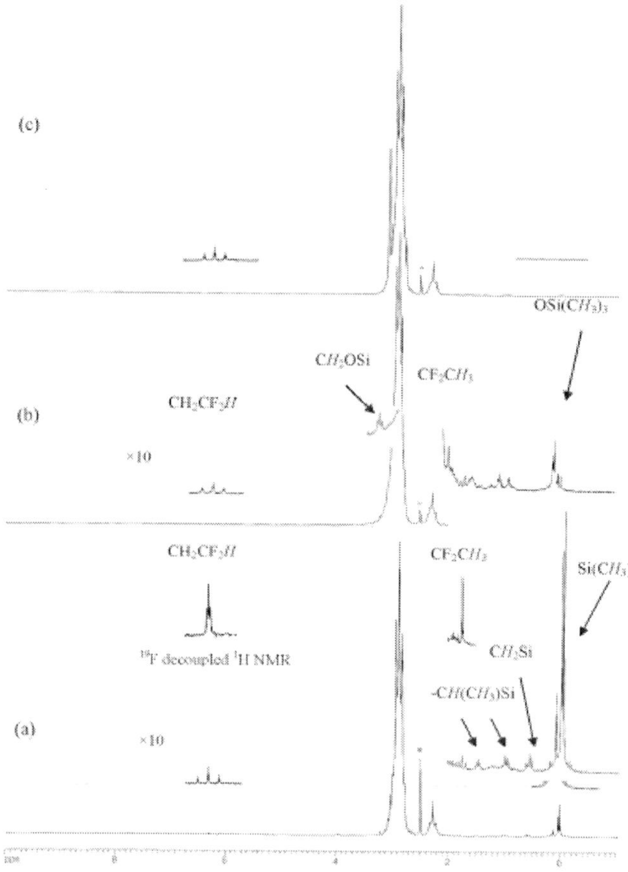


Figure 5. $^1$H NMR spectra of three chain end functionalized PVDF polymers, including (a) PVDF-t-Si$Me_3$ (run H-2), (b) PVDF-t-O-Si$Me_3$ (run I-4) and (c) its corresponding PVDF-t-OH. [Reproduced with permission from Macromolecules 2007, 40, 5235-5240. Copyright 2007 Am. Chem. Soc.].

Figure 5 (a) shows the $^1H$ NMR spectrum of a PVDF-t-$SiMe_3$ polymer (run H-2), with insets showing the expanded regions and chemical shift assignments that are associated with three terminal groups (-$Si(CH_3)_3$, -$CH_2CF_2H$, and -$CF_2CH_3$). Two selected $^{19}F$ decoupled chemical shifts (changing from a triplet-triplet to a triplet peak at 6.3 ppm, and from a triplet to a singlet peak at 1.8 ppm, respectively) confirm the existence of two end group structures (-$CH_2CF_2H$, and -$CF_2CH_3$). Pianca [17] assigned these two groups as the side chain end groups, formed by intra-chain chain transfer reactions. These transfer reactions involve -$CH_2CF_2$* (major), and -$CF_2CH_2$* (minor) propagating sites with a 1,5-hydrogne shift after a head-to-head monomer addition. As discussed in Table 2, the PVDF-t-$SiMe_3$ polymer molecular weight, estimated from the concentration of two $Si(CH_3)_3$ chain end groups (determined by $^1H$ NMR), is similar to that of the GPC result, indicating the existence of the telechelic PVDF polymer structure with two terminal silane groups.

Figures 5 (b) and (c) show the $^1H$ NMR spectra of the end functional PVDF polymers, containing O-$SiMe_3$ (run I-4) and its corresponding OH group (after hydrolysis). The PVDF-t-O-$SiMe_3$ polymer is quantitatively hydrolyzed to form the desirable hydroxyl group terminated PVDF (PVDF-t-OH). Essentially, the same reaction mechanism was observed in each reaction with the incorporation of a functional group in the beginning of the polymer chain. The free radical coupling reaction (termination) took place after monomer-removal in order to form the polymer with two terminal functional groups. Similar results were also observed in the copolymers and terpolymers, which are discussed in the supplemental information.

## 10.4. PP/Clay and PVDF/Clay Nanocomposites

The availability of a broad range of well-defined functionalized PP polymers provides us with a great advantage in evaluating their applications in PP/clay nanocomposites[18]. The most desirable ammonium group terminated i-PP (PP-t-$NH_3^+Cl^-$) polymers were prepared by a simple mixing between PP-t-$NH_2$ and HCl. Both pristine $Na^+$-montmorillonite clay ($Na^+$-mmt), with an ion-exchange capacity of *ca.* 0.95 meq/g, and a dioctadecylammonium-modified montmorillonite organophilic clay (2C18-mmt) were obtained from Southern Clay Products. Static melt intercalation was employed to prepare all PP/clay nanocomposites.

Figure 6 compares the X-Ray Diffraction (XRD) patterns before and after static annealing of a physical mixture (90/10 weight ratio) of an ammonium terminated i-PP (PP-t-$NH_3^+Cl^-$; $M_n$= 58,900 and $M_w$= 135,500 g/mol; $T_m$= 158.2° C) and a pristine $Na^+$-mmt clay. Namely, simple mixing of dried PP-t-$NH_3^+Cl^-$ powder and $Na^+$-mmt—ground together by a mortar and pestle at ambient temperature—creates the XRD pattern in Figure 6(a), with a (001) peak at $2\theta \cong 7^\circ$, corresponding to the characteristic $Na^+$-mmt d-spacing of *ca.* 1.26nm. The mixed powder was then heated (annealed under static conditions) at 190° C for 2 hours under vacuum. The resulting PP-t-$NH_3^+$/mmt hybrid shows a featureless XRD pattern in Figure 6(b), indicating the formation of an exfoliated clay structure. This corresponds to the thermodynamically stable state, as the ammonium-terminated PP exchange the alkali ($Na^+$) cations at the mmt surfaces.

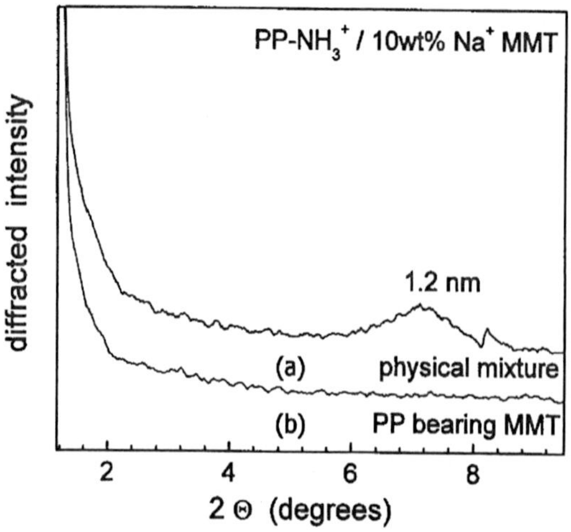


Figure 6. X-ray diffraction patterns of PP-t-$NH_3$+Cl-/Na+-mmt (90/10 weight ratio) (a) physical mixture by simple powder mixing at ambient temperature and (b) the same mixture after static melt-intercalation (PP-t-$NH_3$+/mmt hybrid). [Reproduced with permission from Macromolecules 2003, 36, 8919-8922. Copyright 2003 Am. Chem. Soc.].

Similar featureless patterns were also observed in the PP-t-$NH_3^+Cl^-$/$2C_{18}H_{37}$-$2CH_3$-$N^+$-mmt nanocomposite cases, suggesting that the dioctadecylammonium cations in organophilic clay may also be ion-exchanged by PP-t-$NH_3^+$ to form a similar PP-t-$NH_3^+$/mmt structure. It is clear that an organic surfactant is not needed to promote compatibility between PP-t-$NH_3^+Cl^-$ and pristine $Na^+$-mmt clay. Beyond any economic benefits, such an elimination of the mmt's organic surfactant, this also offers some significant material-related advantages. For example, it eliminates two major concerns relating to the thermal stability of the surfactant during high temperature melt processing and to the long term stability of the small organic surfactant in the polymer/clay nanocomposite under various application conditions. Similar results were also observed in other chain functionalized polyolefin cases. The featureless XRD pattern of the melt blending sample clearly shows the exfoliated PP-t-OH/2C18-mmt nanocomposite structure, implying that PP-t-OH is also a good interfacial agent for exfoliating the clay interlayers.

The binary PP-t-$NH_3^+$/mmt hybrid was subsequently further mixed/blended (at a 50/50 weight ratio) with a neat i-PP (Mn= 110,000 and Mw= 250,000 g/mol). Figure 7 shows the XRD patterns of (a) the physical mixture and (b) the melt blending, of the exfoliated PP-t-$NH_3^+$/mmt structures with neat i-PP. The exfoliated structure is maintained after further mixing with i-PP, which is compatible (co-crystallizable) with the backbone of the largely-isotactic PP-t-$NH_3^+$ polymer, as also directly observed in the TEM image (inset in Figure 7) . Apparently, the i-PP polymer chains largely serve as diluents in the ternary PP-t-$NH_3^+$/mmt/i-PP system, with the thermodynamic stable PP-t-$NH_3^+$/mmt exfoliated structure dispersed in the i-PP matrix [19].

For comparison, several functionalized PP polymers, containing randomly distributed functional groups in the side chains or lumped together in a block-copolymer microstructure, were also evaluated in PP/montmorillonite nanocomposites. Similar static melt-intercalation procedures were followed, except for employing alkylammonium-modified montmorillonites

(2C18-mmt for all random copolymers, and C18-mmt for the block copolymer). Figure 8 shows the XRD patterns of four nanocomposites made with 6 wt.% of 2C18-mmt clay and 94 wt.% of three side-chain functionalized PPs ($M_n$= 95,000 and $M_w$ = 210,000 g/mole) containing (a) 1 mole% p-methylstyrene, (b) 0.5 mole% maleic anhydride, and (c) 0.5 mole% hydroxy side groups, respectively, and the original 2C18-mmt clay. All the functionalized PPs were derived from the same random PP copolymer synthesized by metallocene catalysis, which contained 1 mole% p-methylstyrene (p-MS) comonomer. Subsequently, the p-MS's were selectively functionalized towards hydroxy (OH) and maleic anhydride, respectively, without changing the PP backbone. In addition, we also compared the XRD pattern of PP-*b*-PMMA block copolymer/6 wt% C18-mmt which contains a 5 mole% methyl-methacrylate block These XRD patterns clearly show that there is a definite intercalated structure for all the side chain functionalized PP cases, manifesting itself through an interlayer d-spacing increase of about 1 nm compared to that of the parent alkyl-ammonium-mmt. TEM was also employed to examine the degree of exfoliated (disordered) structure, and the quantitative image analysis indicates only between 20 and 40% exfoliated layers in all four systems [20].

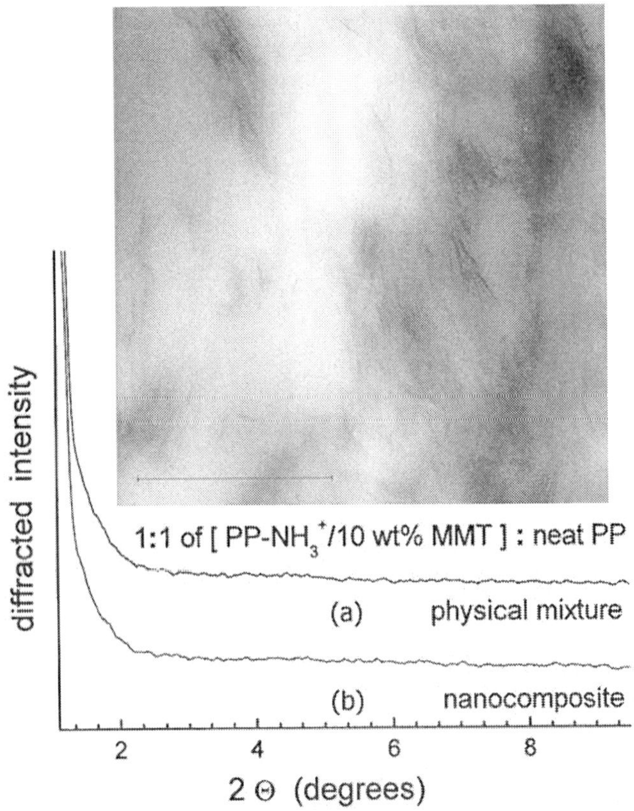


Figure 7. X-ray diffraction patterns of the 50/50 mixture by weight of exfoliated PP-t-$NH_3$+/ mmt structure (90/10 weight ratio) and neat –unfunctionalized– i-PP. The XRD traces shown correspond to (a) the physical mixture of PP-t-$NH_3$+/mmt and i-PP, and (b) the same mixture after static melt-intercalation. Bright-field TEM image of the melt-mixed ternary PP-t-$NH_3$+/mmt/PP system (b in the XRD) shows the exfoliated montmorillonite structure. [Reproduced with permission from Macromolecules 2003, 36, 8919-8922. Copyright 2003 Am. Chem. Soc.].

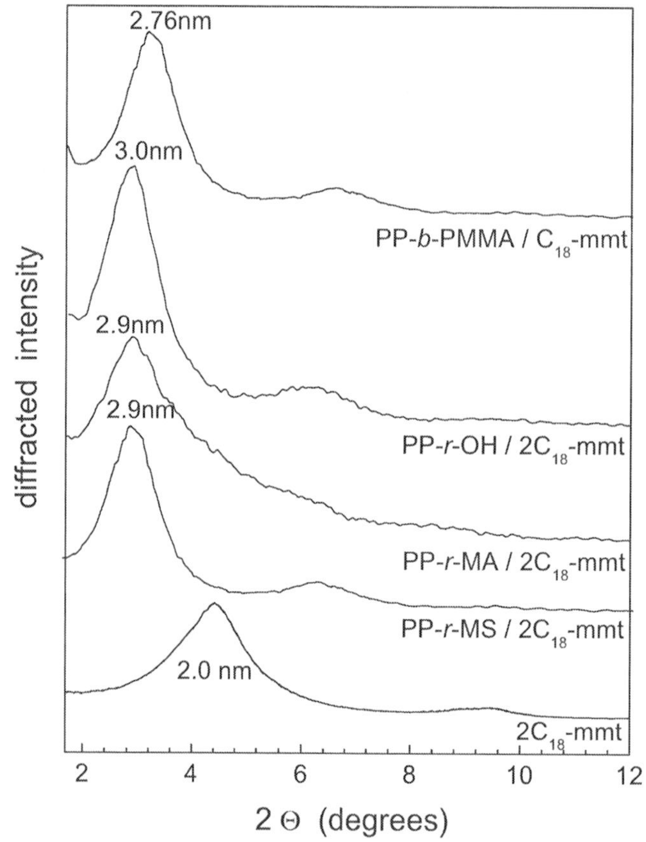


Figure 8. X-ray diffraction patterns of 2C18-mmt clay and four nanocomposites with 6wt.% of alkyl-ammonium-mmt, and 94 wt.% of three side-chain functionalized PPs containing (a) 1 mole% p-methylstyrene, PP-r-MS; (b) 0.5 mole% maleic anhydride, PP-r-MA; and (c) 0.5 mole% hydroxy, PP-r-OH. Also shown, a 6 wt.% C18-mmt nanocomposite of a PP-b-PMMA block copolymer, with 5 mole% of methyl-methacrylate. [Reproduced with permission from Macromolecules 2003, 36, 8919-8922. Copyright 2003 Am. Chem. Soc.].

Resembling the chain-end functionalized polypropylene, the chain-end functionalized fluoropolymer shows an effective dispersion (stabilization) of inorganic particles in the polymer matrix. Figure 9 shows the X-ray results of one set of PVDF/clay samples, in which the PVDF-t-Si(OEt)$Me_2$ containing the reactive Si(OEt)$Me_2$ terminal group (run I-3; Tm = 170 °C, $M_n$ = 30,000 g/mol) was mixed with pristine $Na^+$-mmt clay (without treatment with organic surfactants or acids). Static melt intercalation was employed by, first, mixing and grinding PVDF-t-Si(OEt)$Me_2$ dried powder and $Na^+$-mmt with a 90/10 weight ratio in a mortar and pestle at ambient temperature. The XRD pattern (Figure 9,a) of this simple mixture shows a (001) peak at $2\theta \sim 7$, corresponding to the $Na^+$-mmt interlayer structure with a d-spacing of 1.45 nm. The mixed powder was then heated at 200°C for 3 hours under a nitrogen condition. The resulting PVDF-t-Si(OEt)$Me_2$/$Na^+$-mmt nanocomposite shows a featureless XRD pattern (Figure 9,b), indicating the formation of an exfoliated clay structure. Evidently, the reactive Si(OEt)$Me_2$ terminal group may react with the Si-OH group on the clay surfaces to form a Si-O-Si bond at the interface between clay interlayers. The anchored hydrophobic and oleophobic fluoropolymer chains-disliking the hydrophilic clay surfaces-exfoliate the clay layer structure.

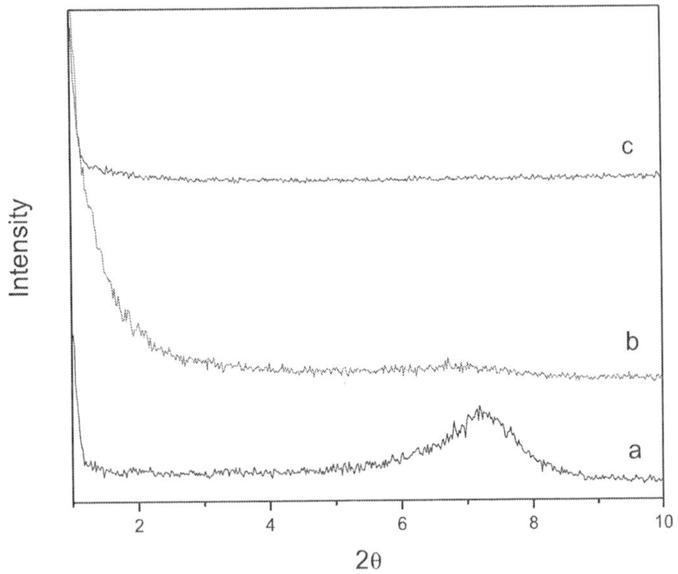


Figure 9. X-ray diffraction patterns of (a) physical mixture of PVDF-t-Si(OEt)$Me_2$ and pristine Na+-mmt clay (90/10 weight ratio), (b) the same mixture after static melt-intercalation, and (c) the 50/50 mixture by weight of exfoliated PVDF-t-Si(OEt)$Me_2$/ Na+-mmt structure (from b) and neat PVDF. [Reproduced with permission from Macromolecules 2007, 40, 5235-5240. Copyright 2007 Am. Chem. Soc.].

The resulting binary PVDF-t-Si(OEt)$Me_2$/$Na^+$-mmt exfoliated nanocomposite was further melted by mixing (50/50 weight ratio) with a commercial neat PVDF (Mn= 70,000 and Mw = 180,000 g/mol). First, the PVDF-t-Si(OEt)$Me_2$/$Na^+$-mmt exfoliated nanocomposite and neat PVDF, with a 50/50 weight ratio, were ground together in a mortar and pestle at ambient temperature. The mixed powder was then heated at 200° C for 3 hours under a nitrogen condition. The resulting ternary PVDF/PVDF-t-Si(OEt)$Me_2$/$Na^+$-mmt nanocomposite also shows a featureless XRD pattern (Figure 9,c), indicating that the stable exfoliated structure in the binary PVDF-t-Si(OEt)$Me_2$/$Na^+$-mmt exfoliated nanocomposite is maintained after further mixing with PVDF that is compatible with the backbone of PVDF-t-Si(OEt)$Me_2$.

## 10.5. Experimental Section

$Me_2$Si(2-methyl-4-phenyl-indenyl)$_2$ $ZrCl_2$ (Boulder Scientific Co.) was used as received. MAO (Albemarle) in 30 wt % of solution in toluene was isolated by vacuum removal of toluene and re-compounded to 2.5 M toluene solution prior to use. CP grade propylene gas (MG Industries) was dried over $CaH_2$ prior to use. CP grade toluene, hexane, and THF were deoxygenated by argon purge before refluxing for 48 hours and then distilled over sodium benzophenone right before use.

$^1$H NMR was measured on a Bruker AM-300 or WM-360 spectrometer with DISNMR software. The measurements were usually taken at 120°C using $d_2$-1,1,2,2-tetrachloroethane as solvent. The polymer molecular weights were determined by Gel Permeation Chromatography (GPC) using a Waters 150 C with a refractive index (RI) detector. The measurements were taken at 140° C using 1,2,4-trichlorobenzene (TCB) as solvent and a mobile phase of 0.7 ml/min flow rate. Broad molecular weight PE samples were used as standards for calibration. Differential scanning calorimetry (DSC) was measured on a Perkin Elmer DSC-7 instrument controller. The DSC curves were recorded during the second heating cycle from 30° to 180°C with a heating rate of 20°C/min. XRD data were collected on a diffractometer using Cu K$\alpha$ radiation ($\lambda$ = 0.1506 nm). Bragg's law ($\lambda = 2d \sin\theta$) was used to calculate the spacing. Direct observation of the PP/clay nanocomposite structure was realized by bright field TEM of nanocomposite films under strain in a JEOL –1200EX operating at 120kV. Ultrathin sections of the PP/clay with a thickness of approximately 50 nm were prepared at room temperature using an ultramicrotome equipped with a diamond knife. The sections were transferred dry to carbon-coated Cu grids of 200 mesh.

### *Synthesis of 4-(t-butyldimethylsilyloxy)styrene (St-OSi)*

4-(*t*-Butyldimethylsilyloxy)styrene was synthesized in two steps. In a 500 ml flask equipped with a magnetic stirring bar, 70.4g (1.02mol) of imidazole was mixed with 52.4g (0.42 mol) of 4-hydroxybenzaldehyde and 77.4g (0.52mol) of *t*-butyldimethylsilyl chloride in THF solution. The mixture was stirred at ambient temperature for 4 hours before being poured into cold water. The organic layer was separated and extracted with ether, then dried with magnesium sulfate. About 94g of 4-(*t*-butyldimethylsilyloxy)benzaldehyde (90% yield) was obtained after evaporating the solvent. The second reaction step was performed under a nitrogen atmosphere. In a 500 ml flask equipped with a magnetic stirring bar, 123.6g (0.38mol) of methyltriphenylphosphonium bromide suspended in THF was treated with 149.6 ml (0.19mol) of *n*-butyllithium(2.5M in hexane). After 1 hour, 80.0g (0.34mol) of 4-(*t*-butyldimethylsilyloxy)benzaldehyde was introduced dropwise into the red solution. The mixture was stirred overnight at room temperature and then poured into cold water. The organic layer was separated by ether extraction and dried with magnesium sulfate. Further purification was performed by distillation under vacuum (10 Torr) at elevated temperature (80°C). 65g of 4-(*t*-butyldimethylsilyloxy)styrene was obtained with a yield of more than 90%.

### *Synthesis of 4-{2-[N,N-Bis(trimethylsilyl)amino]ethyl}styrene (p-NSi$_2$-St)*

A silane protected chain transfer agent p-NSi$_2$-St was prepared in two steps. In a 500 ml flask equipped with a magnetic stirring bar, 100 g of lithium bis(trimethylsilyl)amide dissolved in 200 ml of THF was slowly added into a mixture of 50 ml (0.658 mol) of chloromethyl methyl ether and 50 ml of THF at 0°C under a nitrogen atmosphere. After completing the addition, the solution was allowed to warm up to room temperature for 2 hours before evaporating the excess chloromethyl methyl ether and THF solvent. *N,N*-Bis(trimethylsilys)methoxymethylamine (80% yield) was isolated by distillation. In the second step, p-$NSi_2$-St was prepared by treating 4-vinylbenzylmagnesium chloride with *N,N*-bis(trimethylsilys)methoxymethylamine. In a 500 ml flask equipped with a magnetic stirring bar and a condenser, 15.2 g of magnesium was suspended in 50 ml dry ether, and 80 ml of 4-vinylbenzyl chloride diluted with 50 ml dry ether was then introduced dropwise through the condenser. The solution was refluxed for 4 hours before the addition of 117 g of *N,N*-bis(trimethylsilys)methoxymethylamine over a period of 2 hours. The reaction was allowed to proceed at room temperature for another 2 hours before adding 100 ml of aqueous NaOH solution (30%). The organic layer was separated and dried, and the residual was then distilled over $CaH_2$ to obtain p-$NSi_2$-St with 70% yield.

### *Synthesis of $NH_2$ group terminated PP (PP-t-$NH_2$)*

The p-$NSi_2$-St terminated PP polymers (PP-t-St-$NSi_2$) were prepared by the combination of rac-$Me_2Si[2\text{-}Me\text{-}4\text{-}Ph(Ind)]_2ZrCl_2$/MAO catalyst and p-$NSi_2$-St and $H_2$ as the chain transfer agents. A systematic study was conducted to evaluate the effects of p-$NSi_2$-St and $H_2$ concentrations on the catalyst activity and polymer molecular weight. The PP-t-$NH_2$ polymers were then prepared from PP-t-St-$NSi_2$ polymers by treating them with hydrogen chloride, which can be accomplished during the sample work-up step. Alternatively, the isolated PP-t-St-$NSi_2$ (2 g) was suspended in 50 ml of THF at 50°C before adding dropwise 2 N methanolic hydrogen chloride solution. The mixture was stirred for 4 hours at 50°C, and then poured into 1 N methanolic NaOH solution. The PP-t-$NH_2$ polymer was collected by filtration and washed with 1 M aqueous ammonia and water under a nitrogen atmosphere. The polymer was dried overnight at 50°C under vacuum. The PP-t-$NH_2$ polymer yield was quantitative. Overall, the PP-t-$NH_2$ molecular weight is governed by the chain transfer agent - the higher the concentration of the p-$NSi_2$-St, the lower the molecular weight of the resulting polymer. It is clear that the chain transfer reaction to p-$NSi_2$-St (with rate constant $k_{tr}$) is the dominant termination process, and that it competes with the propagating reaction (with rate constant $k_p$). The degree of polymerization ($X_n$) follows a simple comparative equation $X_n=k_p[\text{olefin}]/k_{tr}[\text{St-NSi}_2]$ with a chain transfer constant of $k_{tr}/k_p=1/34$. $^1H$ NMR spectra of PP-t-St-$NSi_2$ polymer and the corresponding PP-t-St-$NH_2$ show clean deprotection reaction with the complete disappearance of the silane protecting group at 0.24 ppm.

### *Synthesis of $[(C_2H_5O)_3SiCH_2CH_2]_3B$ functional initiator*

In a 500 mL flame-dried flask, equipped with a magnetic stir bar, 250 mL of dry THF and 35 g (180 mmol) of vinyltriethoxylsilane were injected under argon. After cooling the solution to 0°C, 60 mL of $BH_3$ in THF (1.0 M) was added. The mixture was stirred at 0°C for 4 hours and then was warmed to ambient temperature for 1 hour to assure a complete hydroboration reaction. After solvent-removal, the product was subjected to vacuum distillation at 170°C to obtain 23.4 g of the tri-(triethoxylethylsilyl)borane product. The $^1H$ NMR spectrum indicates that the hydroboration reaction involves, mainly, an anti-Markovnikov addition (>90%).

### *Synthesis of PVDF polymers with A terminal $(C_2H_5O)_3Si$ Group*

In a typical reaction (run A-2 in Table 1), 0.6 g of $[(C_2H_5O)_3SiCH_2CH_2]_3B$ (1 mmol) was dissolved in 30 mL of $CH_3CN$ in a dry box. The reactor was then connected to a vacuum line, and 20 mL (13.4 g) of VDF (210 mmol) was condensed into an autoclave reactor under vacuum by liquid nitrogen. VDF has a vapor pressure of ~ 40 atm at 25°C. About 25 mL $O_2$ was charged into the reactor to oxidize borane moiety and initiate the polymerization that was carried out at ambient temperature for 5 hours. After releasing the pressure, the mixture was transferred into a flask containing 100 mL of hexane. After stirring for 30 minutes, the polymer powder was filtered, washed, and then dried under vacuum at ambient temperature for 6 hours. About 4.1 g of polymer was obtained with a yield of 30.6 %.

### *Preparation of PP/PP-t-$NH_3^+Cl^-$/ $Na^+$-mmt nanocomposite*

$Na^+$-montmorillonite clay ($Na^+$-mmt) with an ion-exchange capacity of ca. 95 mequiv/100g (WM) was obtained from Southern Clay Product. PP-t-$NH_3^+Cl^-$ ($T_m$=158.2°C; $M_n$= 58,900 and $M_w$=135,500 g/mole) was prepared using excess HCl reagent. Static melt intercalation was employed to prepare PP-t-$NH_3^+Cl^-$/ $Na^+$-montmorillonite nanocomposite. PP-t-$NH_3^+Cl^-$ dried powder and $Na^+$-mmt with 90/10 weight ratio were first mixed and ground together in a mortar and pestle at ambient temperature. The XRD pattern of this simple mixture shows a (001) peak at 2θ ~7, corresponding to $Na^+$-mmt interlayer structure with a d-spacing of 1.45 nm. The mixed powder was then heated at 190°C for 2 hours under nitrogen condition. The resulting PP-t-$NH_3^+Cl^-$/ $Na^+$-mmt nanocomposite shows a featureless XRD pattern, indicating the formation of an exfoliated clay structure.

The binary PP-t-$NH_3^+Cl^-$/Na-mmt exfoliated nanocomposite was further melt mixed (50/50 weight ratio) with commercial neat i-PP (Mn= 110,000 and Mw= 250,000 g/mol). First, the PP-t-$NH_3^+Cl^-$/Na-mmt exfoliated nanocomposite and neat i-PP with 50/50 weight ratio were ground together in a mortar and pestle at ambient temperature. This simple mixture shows a featureless XRD pattern. The mixed powder was then heated at 190°C for 2 hours under nitrogen condition. The resulting ternary PP/PP-t-$NH_3^+Cl^-$/ $Na^+$-mmt nanocomposite also shows a featureless XRD pattern, indicating that the stable exfoliated structure in the binary PP-t-$NH_3^+Cl^-$/Na-mmt exfoliated nanocomposite is clearly maintained after further mixing with PP that is compatible with the backbone of PP-t-$NH_3^+Cl^-$.

### *Preparation of PVDF/ PVDF-t-Si(OR)$_3$/ $Na^+$-mmt nanocomposite*

The silane terminated PVDF polymers were used as interfacial agents in the preparation of exfoliated PVDF/clay nanocomposites by a melt blending process. In a typical example, a PVDF-t-Si(OR)$_3$ polymer (Tm= 168°C, Mn= 22,100 g/mol) was mixed with pristine $Na^+$-mmt clay, which has an ion-exchange capacity of ca. 90 mequiv/100g. Static melt intercalation was employed by, first, mixing and grinding PVDF-t-Si(OR)$_3$ dried powder and $Na^+$-mmt with a 90/10 weight ratio in a mortar and pestle at ambient temperature. The mixed powder was then heated at 200°C for 3 hours under nitrogen to result in the exfoliated PVDF-t-Si(OR)$_3$/$Na^+$-mmt nanocomposite. The resulting binary PVDF-t-Si(OR)$_3$/$Na^+$-mmt exfoliated nanocomposite was further melted by mixing it (50/50 weight ratio) with commercial neat PVDF (Mn= 70,000 and Mw = 180,000 g/mol). First, the PVDF-t-Si(OR)$_3$/$Na^+$-mmt exfoliated nanocomposite and neat PVDF with 50/50 weight ratio were ground together in a mortar and pestle at ambient temperature. The mixed powder was then heated at 200°C for 3 hours under nitrogen to form a ternary PVDF/ PVDF-t-Si(OR)$_3$/$Na^+$-mmt nanocomposite.

# 10.6. Conclusions

In this chapter, I have summarized our experimental results on the preparation of chain-end functionalized PP and PVDF polymers that serve as the interfacial agent in forming the exfoliated polymer/clay nanocomposites. A one-pot process to prepare chain-end functionalized PP polymers involves a metallocene-mediated propylene polymerization/chain transfer reaction. With the proper choice of reaction conditions, chain transfer agents, and catalyst system, it is very convenient and efficient (with high catalyst activity) to prepare i-PP with a terminal styrenic unit containing either an OH, or $NH_2$ group. This reaction scheme is especially useful in the preparation of chain-end functionalized PP with high molecular weight. Almost the same reaction procedures used in regular PP polymerization can be directly applied, with the exception of adding a small amount of a chain transfer agent. On the other hand, we have also developed a new synthesis route through the combination of functional borane initiators, control radical polymerization, and termination via the radical coupling reaction, which produces telechelic fluoropolymers containing a terminal functional group. This chemistry is advantaged by its simplicity and mild reaction condition, and is applicable to a broad range of fluoromonomers, including VDF, TrFE, CTFE, HFP, etc.

Appending such benefits, the formed chain-end functionalized polymers, including PP and PVDF, are very effective polymeric surfactants, which shows high interfacial activity in the exfoliated polyolefin/clay nanocomposites. As illustrated in Figure 10(a), the functional group anchors the polymer chain to the clay interface. The rest of the unperturbed high molecular weight hydrophobic or/and oleophobic polymer chain-disliking the hydrophilic clay surfaces-exfoliate the clay layer structure. This disordered clay structure is maintained even after further mixing with a neat (unfunctionalized) polymer that is compatible with the backbone of the chain-end functionalized polymer. In contrast, side-chain functionalized or block copolymers form multiple contacts with each of the clay surfaces, as illustrated in Figure 10(b). The results not only in aligned polymer chains parallel to the clay surfaces, but also results in bridged consecutive clay platelets promoting intercalated structures, especially for the higher lateral size montmorillonites.

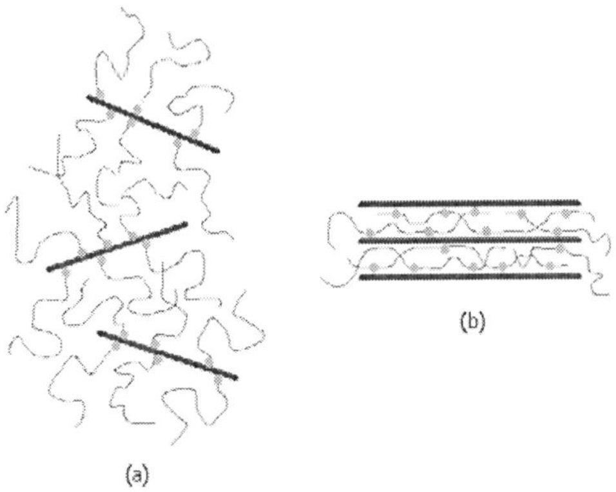


Figure 10. Illustration of the molecular structures of (a) chain end functionalized and (b) side chain functionalized polyolefin located between clay interlayers. [Reproduced with permission from Macromolecules 2003, 36, 8919-8922. Copyright 2003 Am. Chem. Soc.].

## Acknowledgment

The author would like to thank the Office of Naval Research for the financial support.

## References

Chung, TC. *Functionalization of Polyolefins;* Academic Press: London, 2002. (b) Chung, TC. *Prog. Polym. Sci,* 2002, 27, 39.

Chung, TC. *ChemTech,* 1991, 27, 496. (b) Chung, TC; Rhubright, D. *Macromolecules,* 1991, 24, 970.

Usuki, A; Kojima, Y; Kawasumi, M; Okada, A; Fukushima, Y; Kurauchi, T; Kamigaito. O. *J. Mater. Res.,* 1993, 8, 1179. (b) Kojima, Y; Usuki, A; Kawasumi, M; Okada, A; Kurauchi, T; Kamigaito, O. *J. Polym Sci., Part A: Polym. Chem,* 1993, 31, 983. (c) Vaia, RA; Ishii, H; Giannelis, EP. *Chem. Mater.,* 1993, 5, 1694. (d) Giannelis, EP; Krishnamoorti R, Manias E. *Adv. Polym. Sci,* 1999, 138, 107.

Balazs, AC; Singh, C; Zhulina, E. *Macromolecules,* 1998, 31, 8370; Kuznetsov D. V; Balazs, AC. *J. Chem. Phys*, 2000, 113, 2479. (b) Vaia RA; Giannelis EP. *Macromolecules*, 1997, 30, 7990.

Beyer, FL; Tan, NCB; Dasgupta, A; Galvin, ME. *Chem. Mater.,* 2002, 14, 2983. (b) Huang, XY; Brittain, W. J. *Macromolecules* 2000, 34, 3255; Okamoto, M; Morita, S; Taguchi, H; Kim, YH; Kotaka, T; Tateyama, H. *Polymer* 2000, 41, 3887. (c) Messersmith, PB; Giannelis, EP. *J. Polym. Sci., Part A: Polym. Chem,* 1995, 33, 1047.

Napper, D. *Polymeric Stabilization of Colloidal Dispersion;* Academic Press: London, 1983. (b) Gast, A; Leibler, L. *Macromolecules,* 1986, 19, 686. (c) Lee, LH. *Adhesion and Adsorption of Polymers;* Plenum Press: New York, 1980. (d) Koning, C; Van Duin, M; Pagnoulle, C; Jerome, R. *Prog. Polym. Sci.*, 1998, 23, 707. (e) Noolandi, J; Hong, K. *Macromolecules* 1982, 15, 482. (f) Milner, ST; Xi, H. *J. Rheol.*, 1996, 40, 663.

Bamford, CH; Jenkins, AD; Wayne, RP. *Trans. Faraday Soc.,* 1960, 56, 932. (b) Haas, HC; Schuler, NW; Kolesinski, HS. *J. Polym. Sci, Part A1,* 1967, 5, 2964.

Brody, H; Richards, DH; Szwarc, M. *Chem. Ind.*, 1958, 1473. (b) Hayashi, K; Marvel, CS. *J. Polym. Sci., Part A1,* 1974, 2, 2571. (c) Morton, M; Fetters, LJ; Inomater, J; Rubio, DC; Young, RN. *Rubber Chem. Technol,* 1976, 49, 303.

Yasuda, H; Furo, M; Yamamoto, H; Nakamura, A; Miyake, S; Kibino, N. *Macromolecules* 1992, 25, 5115. (b) Brookhart, M; Desimone, JM; Grant, BE; Tanner, MJ. *Macromolecules,* 1995, 28, 5378. (c) Shea, KJ; Walker, JW; Zhu, H; Paz, M; Greaves, J. *J. Am. Chem. Soc.,* 1997, 911, 9049. (d) Desurmont, G; Tokimitsu, T; Yasuda, H. *Macromolecules,* 2000, 33, 7679.

Xu, G; Chung, TC. *J. Am. Chem. Soc.,* 1999, 121, 6763. (b) Xu, G; Chung, TC. *Macromolecules,* 1999, 32, 8689. (c) Chung, TC; Xu, G; Lu, Y; Hu, Y. *Macromolecules,* 2001, 34, 8040.

Chung, TC; Dong, JY. *J. Am. Chem. Soc.,* 2001, 123, 4871. (b) Dong, JY; Chung, TC. *Macromolecules,* 2002, 35, 1622. (c) Dong, JY; Han, H; Chung, TC. *Macromolecules,* 2002, 35, 9352. (d) Chung, TC; Dong, JY. *U. S. Patent,* 6,479,600 (2002).

Chung, TC; Lu, HL; Janvikul, W. *J. Am. Chem. Soc,* 1996, 118, 705. (b) Hong, H; Chung, TC. *Macromolecules,* 2004, 37, 6260. (c) Zhang, ZC; Chung, TC. *Macromolecules,* 2006, 39, 5187.

Chung, TC; Petchsuk, A. *Macromolecules,* 2002, 35, 7678. (b) Chung, TC; Petchsuk, AUS. Patent 6,355,749 (2002). (c) Wang, ZM., Zhang, ZC; Chung, TC. *Macromolecules,* 2006, 39, 4268. (d) Zhang, ZC; Chung, TC. *Macromolecules,* 2007, 40, 783.

Hong, H; Chung TC. *Macromolecules,* 2004, 37, 6260. (b) Chung, TC; Hong, H; Oka, M; Kubo, K. *U. S. Patent,* 7,262,257 (2007). (c) Chung, TC; Hong, H. In *Advances in Controlled/Living Radical Polymerization*; Matyjaszewski, K; Ed; ACS Symposium Series No. 854, 2003; pp 481-495. (d) Chung, TC; Hong, H; Zhang, ZC; Wang, ZM. In *Controlled/Living Radical Polymerization*; Matyjaszewski, K; Ed; ACS Symposium Series No. 944, 2006; pp 387-403.

Zhang, ZC; Wang, ZM; Chung, TC. *Macromolecules* 2007, 40, 5235. (b) Zhang, ZC; Chen, S. T; Chung, TC; Kubo, K; Kigawa, KUS. Patent Application Publication No. US2008/0081195 A1.

Brown, HC. *Organic Synthesis via Borane;* Wiley-Interscience: New York, 1975. (b) Mikhailov, BM; Bubnov, YN. *Organoboron Compounds in Organic Synthesis,* Harwood Academic Publishers: Switzerland, 1984.

Pianca, M; Barchiesi, E; Esposto, G; Radice, S. *J. Fluorine Chem,* 1999, 95, 71.

Chung, TC; Manias, E; Wang, ZM. *U. S. Patent* 7,241,829 (2007). (b) Wang, ZM; Nakajima, H; Manias, E; Chung, TC. *Macromolecules,* 2003, 36, 8919.

Krishnamoorti, R; Giannelis, EP. *Langmuir,* 2001, 17, 1448. (b) Krishnamoorti, R; Ren, JX; Silva, AS. *J. Chem. Phys.,* 2001, 114, 4968. (c) Ren, JX; Casanueva, BF; Mitchell, CA; Krishnamoorti, R. *Macromolecules,* 2003, 36, 4188.

Manias, E; Touny, A; Wu, L; Strawhecker, K; Lu, B; Chung, TC. *Chem. Mater,* 2001, 13, 3516.

# Authors' Affiliations

## Editor

**Dr. Vikas Mittal**, BASF SE, G201, Polymer Research, 67056, Ludwigshafen, Germany.
vikas.mittal@chem.ethz.ch
vikas.mittal@basf.com

## Chapters

**Chapter I. Need of New Surface Modifications**

V Mittal, BASF SE, G201, Polymer Research, 67056, Ludwigshafen, Germany.
vikas.mittal@chem.ethz.ch
vikas.mittal@basf.com

**Chapter II. Exfoliation through Esterification for Clay Polymer Nanocomposites**

BSJ Prakash (corresponding author), YS Bhat and C Ravindra Reddy, Clay Research Laboratory, Bangalore Institute of Technology, K.R. Road, Bangalore-560 004, India.
jprak27@yahoo.com

**Chapter III. Grafting of Polymer Chains 'from' the Clay Surface**

Y Yang and H Zhao (corresponding author), Department of Chemistry, Nankai University, Tianjin 300071, P. R. China.
hyzhao@nankai.edu.cn

**Chapter IV. Role of Monocationic and Bicationic Initiators on the Grafting of Polymer Chains 'from' the Clay Surface**

P Uthirakumar, Department of Semiconductor and Chemical Technology, Semiconductor Physics Research Center, Chonbuk National University, Jeonju, South Korea.
uthirakumar@gmail.com

**Chapter V. Grafting from the clay surface using atom transfer radical polymerization**
DA Shipp, Department of Chemistry & Biomolecular Science, and Center for Advanced Materials Processing, Clarkson University, Potsdam NY 13699-5810, USA.
dshipp@clarkson.edu

**Chapter VI. Exchange of Functional Modifications on the Clay Surface**
J Duchet-Rumeau, Université de Lyon, F-69003, Lyon, France and INSA Lyon, F-69621, Villeurbanne, France, CNRS, UMR5223, Ingénierie des Matériaux Polymères
Jannick.duchet@insa-lyon.fr

**Chapter VII. Physical Adsorption on Clay Surface**
JQ Jiang, Division of Civil, Chemical and Environmental Engineering, C5, Faculty of Engineering and Physical Sciences, University of Surrey, Guildford, Surrey, GU2 7XH, UK.
j.jiang@surrey.ac.uk

**Chapter VIII. Role of Clean Clay Surface on Composite Properties**
V Mittal, BASF SE, G201, Polymer Research, 67056, Ludwigshafen, Germany.
vikas.mittal@chem.ethz.ch
vikas.mittal@basf.com

**Chapter IX. Advances in Surface Functionalization of Carbon Nanotubes**
V Mittal, BASF SE, G201, Polymer Research, 67056, Ludwigshafen, Germany.
vikas.mittal@chem.ethz.ch
vikas.mittal@basf.com

**Chapter X. Exchange of Chain-end Functionalized Polyolefins on the Clay Surface**
TCM Chung, Department of Materials Science and Engineering, The Pennsylvania State University, University Park, PA 16802, USA
chung@ems.psu.edu

# Index

## A

## D

## N

## Q



## R

## S

## T

## U



## V



## W

[redacted]